Micro-irrigation of Trees and Vines:

A Handbook for Water Managers

Larry Schwankl, *UC Irrigation Specialist*
Blaine Hanson, *UC Irrigation and Drainage Specialist*
Terry Prichard, *UC Water Management Specialist*

University *of* **California**
Agriculture and Natural Resources

ANR Publication 3378

To order or obtain ANR publications and other products, visit the ANR Communication Services online catalog at http://anrcatalog.ucdavis.edu or phone 1-800-994-8849. You can also place orders by mail or FAX, or request a printed catalog of our products from

University of California
Agriculture and Natural Resources
Communication Services
1301 S. 46th Street
Building 478 - MC 3580
Richmond, CA 94804-4600
Telephone 1-800-994-8849
510-665-2195
FAX 510-665-3427
E-mail: danrcs@ucdavis.edu

Publication 3378
ISBN-13: 978-1-60107-752-3

Cover design: Ellen Guttadauro
Cover art: Christine Sarason
This publication was originally funded by the California Energy Commission and the U.S. Department of Agriculture Water Quality Initiative

UC PEER REVIEWED This publication has been anonymously peer reviewed for technical accuracy by University of California scientists and other qualified professionals. This review process was managed by the ANR Associate Editor for Land, Air and Water Resources.

POD-2/11

Contents

Operating the System

Maintaining a Micro-irrigation System

Heading Off Problems

Appendix

Preface

Micro-irrigation of Trees and Vines is the sixth in a series of water management handbooks prepared by the University of California Irrigation Program with funding provided by the California Energy Commission and the U.S. Department of Agriculture Water Quality Initiative. These publications are intended to help California water managers address a range of practical irrigation issues. Other titles in the series include: *Agricultural Salinity and Drainage; Surface Irrigation; Surge Irrigation; Irrigation Pumping Plants;* and *Drip Irrigation for Row Crops.* Ordering information appears on the title page reverse of this handbook.

The authors would like to thank Ricardo Amon and Tony Wong of the California Energy Commission staff for their help in making this publication series possible.

Introduction

Micro-irrigation — applying small amounts of water slowly and frequently through emitters spaced along polyethylene tapes or tubing—makes it possible to apply water precisely where it is needed and to apply it with a high degree of uniformity, lessening both surface runoff — excess water running off the lower end of the field — and deep percolation — water flowing down through the soil past the root zone where it can no longer be used by the plant.

Converting from conventional surface irrigation to a micro-irrigation system therefore can greatly improve how evenly water is applied over a field and how efficiently water is used. But this potential can only be realized if the micro-irrigation system is carefully designed, maintained, and managed. This handbook has been developed to provide the information necessary to help water managers achieve that goal. Intended as a practical guide to selecting and operating a micro-irrigation system, the handbook is written so as to be easily understandable to anyone with a general agricultural background. While the book is aimed primarily at micro-irrigation system managers, irrigation system designers and others interested in micro-irrigation may also find the handbook useful.

The information presented here is grounded both in technical research and in our own field experience. While the separate chapters complement one another when taken as a whole, each also stands on its own so that it is not necessary to start at the beginning of the book and read all the way through. We suggest instead that the reader use the table of contents or turn to the chapter "Components and Considerations: An Overview" for help in locating topics of interest.

Please direct any comments or questions to Larry Schwankl, UC Irrigation Specialist; Blaine Hanson, UC Irrigation and Drainage Specialist, Department of Land, Air and Water Resources, University of California, Davis, CA 95616; telephone number: (530) 752-1130; fax number: (530) 752-5262; or Terry Prichard, UC Water Management Specialist, 420 S. Wilson Way, Stockton, CA 95205; telephone number: (209) 468-2085.

Tables

Figures

Why a Micro-irrigation System?

Advantages and Disadvantages

By Larry Schwankl, UC Irrigation Specialist

Micro-irrigation is currently used on approximately 2.5 million acres of irrigated farmland in the U.S. Of this total, approximately 1.5 million acres of micro-irrigated land is in California. It is estimated that micro-irrigation in California is used on: 120,000 acres of deciduous trees, 300,000 acres of sub-tropical trees (citrus, olives, avocados, etc.), and 300,000 acres of grapes. This amounts to approximately 18% of the deciduous tree acreage, 72% of the sub-tropical tree acreage, and 41% of the grape acreage.

Much of the acreage being newly planted or replanted to trees or vines is adopting micro-irrigation as its irrigation system. As land which is not as high of quality, difficult to level, and not supplied by existing irrigation water supply systems is brought into production; micro-irrigation system adoption becomes more common. Growers are also switching from surface or sprinkler irrigation systems to micro-irrigation systems as irrigation districts become more flexible in water delivery schedules. Water supply from surface water sources which is unreliable due to drought and/or changing water policies is also encouraging growers to switch to micro-irrigation systems.

Like surface and sprinkler irrigation, micro-irrigation offers both advantages and disadvantages. A carefully designed micro-irrigation system can make the most of these advantages, often while compensating for the disadvantages as well.

Advantages

• *Improved plant response.* With a micro-irrigation system, irrigation takes place frequently (often daily), enabling the water manager to maintain the soil moisture at an optimum level. A well-designed, well-maintained micro-irrigation system also applies water evenly, ensuring that all parts of the field receive the same amount of water. These features, together with effective irrigation scheduling, can improve plant growth and yield.

• *Increased irrigation efficiency.* Micro-irrigation systems can make irrigation more efficient and therefore may require less applied water than sprinkler or surface irrigation systems — although these other systems can achieve similar levels of efficiency if the water manager can commit the time and resources necessary for optimum irrigation performance.

A micro-irrigation system can improve irrigation efficiency by:

- reducing evaporation from the soil surface, since only a portion of the field surface is wetted;

- reducing or eliminating runoff;

- reducing deep percolation — water passing below the root zone;

- eliminating the need to over-irrigate some parts of the field because of uneven water application

• *Improved chemical application.* Injecting fertilizers and other chemicals through a micro-irrigation system has a number of advantages over conventional chemical application methods. When a micro-irrigation system is used for chemigation or fertigation:

- the chemicals are applied to the soil in areas only where roots are actively growing;

- since irrigations are frequent, chemical applications can be better timed, making it possible to closely match fertilizer delivery with plant nutritional needs;

- because deep percolation is lessened, chemicals are less likely to be lost by moving past the root zone with the water. Potential harm to the environment is therefore reduced.

• *Reduced weed growth.* Since only a portion of the field is wetted, total weed growth may be less, although weed problems may increase in the areas kept continually wet. The continual wetting may accelerate the breakdown of herbicides used to control weeds.

• *Decreased energy requirements.* A micro-irrigation system may require less energy than a conventional system if it increases irrigation efficiency and therefore requires less water to be pumped. Switching from a conventional pressurized system to a micro-irrigation system can reduce energy needs by as much as 50%. But switching to a micro-irrigation system from a surface irrigation system with water delivered through an irrigation district may *increase* energy requirements.

A properly designed and operated system allows the water manager to easily take advantage of lower energy rates offered during off-peak hours, but micro-irrigation systems designed to operate 24 hours a day during peak crop water demand periods (which in any case is not recommended) would not be able to take advantage of off-peak rates.

• *Automation.* A micro-irrigation system can be easily automated using electrical solenoid valves and a controller. This allows the system to be operated any time of the day or night and for any desired length of time, enabling water managers to take advantage of available crop water use information in determining optimum irrigation time. Soil moisture sensing devices such as tensiometers and electrical resistance blocks can be used to further automate micro-irrigation systems.

• *Reduced salinity hazard.* A micro-irrigation system can lessen salinity problems by:

> • causing salts to remain diluted because of the continued optimum soil moisture content resulting from frequent irrigations;

> • moving salts out to the edges of the wetted soil area, leaving a zone of wetted soil with a lowered salt content for root activity;

> • eliminating the opportunity for salts to be absorbed through leaves, as may occur in sprinkler irrigation.

• *Adaptability to difficult soil and terrain conditions.* Micro-irrigation can be used successfully on steep or undulating terrain not suited to any other irrigation system type, on soils with low water infiltration rates, and on soils with low water-holding capacities.

Disadvantages

• *Maintenance Requirements.* The small passageways characteristic of micro-irrigation emitters make micro-irrigation systems particularly susceptible to clogging from particulate matters, organic matter, and chemical precipitates. Incorporating filtration in the system design can lessen clogging, but additional maintenance — including injecting chlorine or acid and routinely flushing lateral lines — may be necessary to ensure top performance.

Machinery, animals, or foot traffic in the field can cause leaks in the polyethylene lateral lines or at the emitters. Some growers make a practice of inspecting the system and correcting any problems before each irrigation.

• *Cost.* The underground PVC pipe, polyethylene tubing, filters, and other hardware required make drip systems among the most expensive of all irrigation systems. But the benefits — making very uniform water application possible and allowing the water manager to control the amount and timing of irrigations — may justify the expense. Growers contemplating micro-irrigation should weigh the costs against benefits of a micro-irrigation system according to the conditions at hand.

• *Restricted root zone.* Particularly in regions of low rainfall, plant root activity is often limited to the soil zone wetted by the micro-irrigation emitters — usually a much smaller soil volume than that wetted by full coverage sprinkler or surface irrigation systems. While crops grown under micro-irrigation systems have been shown to respond well, the micro-irrigation system manager must remember that the system is meant to apply small, frequent irrigations. Unlike crops grown under surface and sprinkler systems, crops grown under micro-irrigation do not have large reserves of water stored in the soil and therefore cannot endure long periods between irrigations. Problems in the micro-irrigation system — especially those occurring during peak water use — must therefore be repaired quickly to prevent water stress damage to the crop.

A rule of thumb: The micro-irrigation system should wet 50 percent or more of the soil volume explored by the crop roots. The lower the water-holding capacity of the soil, the greater the wetted area should be. This amount of wetting provides adequate soil moisture storage and nutrient uptake as well as adequate anchoring for the crop.

• *Salt accumulation near the root zone.* Unlike surface and sprinkler irrigation systems, which can flush salts below the crop root zone, micro-irrigation systems tend to move salts to the outer edge of the wetted volume of soil and soil surface. Careful management is therefore necessary to ensure that the salts do not migrate back into the active root zone. Rainfall, insufficient to totally leach the salts from the root zone, can move the salts into the root zone and cause damage.

If the need to leach salts from the root zone becomes critical, a sprinkler or surface irrigation system may have to be used to accomplish this purpose effectively.

Components and Considerations: An Overview
By Larry Schwankl, UC Irrigation Specialist

The various types of micro-irrigation systems—micro-sprinkler, surface drip, and subsurface drip systems—are all made up of the same basic components. Generally, a micro-irrigation system consists of:

> a pump
> a flowmeter
> valves
> a filter(s)
> injection equipment
> mainlines and submains
> polyethylene lateral lines
> emission devices

For the most part, micro-irrigation systems differ only in the emitter spacing, the type of emission device used, and the size of the components. The type of emitters used affects the size needed for the other components. Micro-sprinklers, for instance, generally require larger filters, mainlines, and submains.

Figure 1 on the following page shows the components of a typical micro-irrigation system.

In a micro-irrigation system for trees and vines, the emitter spacing and discharge rate needed depends primarily on the spacing of the plants and on the water needs of the crop. The emission devices must be capable of supplying each plant with enough water during the peak water use periods to satisfy the evapo-transpiration requirement—the amount of water used by the crop since the last irrigation.

Following is a brief overview of the components and operation of a micro-irrigation system for trees or vines, along with references to more detailed information located elsewhere in this handbook.

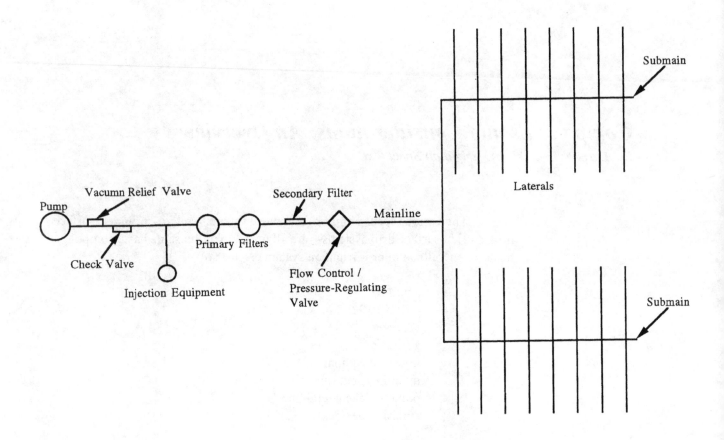

Figure 1. Components of a micro-irrigation system.

Pumping Plant

It is important to select a pump and motor (or engine) that will deliver the correct pressure and flowrate at the highest possible efficiency. The micro-irrigation system designer determines the flowrate and pressure to be delivered by the pump, and the pump dealer uses this information to select the most efficient pump for a given system. *(See "Choosing a Pump," page 35.)*

Filters

Selecting the appropriate filter requires considering water quality factors. Particulate matter (such as sand) in the water can be removed with vortex filters (frequently referred to as "sand separators"). Screen filters are also effective in removing particulate matter. Screen or sand media filters are usually used for removing particulate matter finer than that which can be removed with a sand separator.

Organic matter such as algae or slimes can be removed using either screen or sand media filters, but since organic matter can quickly clog a screen filter and is difficult to flush from the screen, sand media filters are the usual choice for filtering surface waters containing algae and slimes.

Both screen and media filters must be periodically backwashed. The pressure drop across the filter indicates when backwashing is required. Back-washing can be accomplished either manually or automatically, with automatic backwashing taking place either on a defined time schedule or when the filter senses a pre-determined pressure drop across the filter and begins the backwash cycle. The water used to backwash is frequently discharged out of the system. *(See "Filtration Equipment," page 51.)*

Flowmeters

It is important that a flowmeter be part of the irrigation system. Knowing the flowrate is necessary for determining the amount of water being applied, which, in turn, is critical to efficient irrigation and scheduling. A propeller meter, which reports either the flowrate or total water applied, gives an accurate measurement. *(See "Flowmeters", page 41.)*

Injection Equipment

Micro-irrigation is well-suited to injecting chemicals such as chlorine and fertilizer. Various types of injection equipment — differential pressure tanks ("batch tanks"), venturi devices, and electrically driven or water-driven pumps — can be used, depending on the chemical applied, the accuracy level needed, and the injection rate required. *(See "Injection Devices," page 57.)*

Valves

Valves are the control mechanisms of micro-irrigation systems. Several types are common: control valves, air/vacuum relief valves (which allow air to escape when the system is turned on and to enter when the system is shut down), and check valves (which prevent undesirable flow reversal). Pressure-regulating valves are important for maintaining a constant operating pressure in the system. *(See "Valves and Regulators," page 39.)*

Mainlines and Submains

Main and submain pipes, usually made of PVC, deliver water to the lateral lines and emitters. The mains and submains must be sized carefully, with the cost of the pipe balanced against pressure losses caused by friction as water moves through the pipe. The system should be designed by a qualified micro-irrigation system designer. *(See "Calculating Pressure Loss in Mainlines and Submains," page 77.)*

Lateral Lines

Emitters are attached to tubing or lateral lines, usually made of polyethylene. The length and diameter of the lateral lines to be used depends on economics, balancing the tubing cost against pressure loss along the lateral. If the lateral lines are too long or the wrong diameter, the emitters may discharge water at different rates, resulting in non-uniform irrigation. *(See "Calculating Pressure Loss in Lateral Lines," page 79.)*

Emitters

The many different types of micro-irrigation emitters available can be grouped generally into drippers, bubblers, drip tapes, and micro-sprinklers. The appropriate emitter to use may depend on the crop. Row crops are often irrigated by drip tapes placed either on the surface or below the ground. Tree and vine crops are most often irrigated with above-ground drip systems or micro-sprinklers. Micro-sprinklers have become popular for tree crops in recent years, but subsurface drip systems are sometimes used. *(See "Emission Devices," page 25; "Pressure-Compensating Emitters," page 33; "What is an Appropriate*

Lateral Length," page 21; and "Selecting Drip Emitters and Micro-sprinklers," page 29.)

Operation and Maintenance

Micro-irrigation systems can apply irrigation water quite efficiently, but only if they are operated and maintained properly.

Irrigation scheduling — determining when to irrigate and how much water to apply — is critical to operating the system efficiently. Effective scheduling requires knowing how much water the crop is using or has used since the last irrigation (the evapotranspiration or ET of the crop) and how much water the irrigation system applies in a given period of time (the application rate). From that information can be calculated how long the system should run. *(See "Wetting Patterns," page 65; "How Often to Irrigate," page 69; and "How Much Water is Being Applied," page 71.)*

A principal virtue of a micro-irrigation system is its ability to deliver a uniform amount of water to each location it serves so that water is applied evenly over the field. Because of pressure differences through the system and variability in emitter manufacture, even new systems do not apply water completely evenly. A carefully designed system can use pressure regulators or pressure-compensating emitters to overcome pressure differences within the irrigation system.

While a well-designed system can deliver water with a high degree of uniformity, the system must be properly maintained to keep the water application uniform. A principal cause of non-uniformity in micro-irrigation systems is emitters becoming clogged by particulate or organic matter or from lime or iron precipitation. Good filtration, along with adding acid or chlorine to the water when necessary, is therefore crucial to keeping the system running properly. *(See "Routine Maintenance," page 89; "Chlorination," page 93; "Assessing Water Quality," page 97 and "Chemical Precipitate Clogging," page 103.)*

Cost

The initial cost of a micro-irrigation system is approximately $900 to $1300 per acre, which is comparable to that of a solid-set sprinkler system.

Irrigating Efficiently
By Larry Schwankl, UC Irrigation Specialist

Micro-irrigation efficiency is measured in two ways: by application efficiency (AE) and by irrigation efficiency (IE). Application efficiency is normally used to measure a single irrigation event, while irrigation efficiency is used to describe seasonal irrigation water use.

Irrigation efficiency is a measure of the portion of the total applied irrigation water beneficially used — with the principal beneficial use being satisfying crop water needs. Other potential beneficial uses include salt leaching, frost protection, and chemical application.

Irrigation Efficiency

Irrigation efficiency is defined as follows:

$$\text{Irrigation Efficiency (\%)} = \frac{\text{Beneficially used water}}{\text{Total water applied}} \times 100 \qquad (1)$$

Non-beneficial water uses include water lost to deep percolation (that is, water draining below the crop root zone), unused runoff from the field (usually minimal in micro-irrigation systems), and evaporation from wet soil surfaces. The potential irrigation efficiency of a well-operated and well-maintained micro-irrigation system is 75 to 85%.

Application Efficiency

Application efficiency describes an irrigation event as follows:

$$\text{Application Efficiency (\%)} = \frac{\text{Water stored in the crop root zone}}{\text{Total water applied}} \times 100 \qquad (2)$$

Application efficiency takes into account irrigation water lost to deep percolation and unused runoff, but does not take into account beneficial uses other than satisfying crop water needs.

The keys to irrigating efficiently are knowing how much water the crop has used since the last irrigation (net irrigation amount) and operating the irrigation system to apply the amount of water needed. The irrigation amount needed (referred to as the "gross irrigation") allows not only for the water used by the crop, but also for inefficiency and non-uniformity in the irrigation system.

The relationship between the gross irrigation amount and application efficiency is as follows:

$$\text{Gross irrigation amount} = \frac{\text{Net irrigation amount}}{\text{Application efficiency (\%)}} \times 100 \quad (3)$$

The gross irrigation amount is greater than the net irrigation amount because it allows for the inefficiencies in the irrigation.

Crop Water Use

The amount of water used by the crop since the last irrigation can be estimated by keeping track of the crop's water use (or *evapotranspiration,* which is the amount of water transpired by the plant and evaporated from the soil in a given period) or by measuring the soil moisture depletion using such devices as tensiometers, resistance blocks, neutron probes, or dielectric constant meters. Reference evapotranspiration (ETo) and crop water use estimates are available from a number of sources, including the California Department of Water Resources' CIMIS program and University of California Cooperative Extension offices.

Evapotranspiration data is available both as historical, long-term averages and as real-time estimates. Historical averages are useful in advance irrigation planning, but real-time evapotranspiration information is more accurate. A micro-irrigation system provides the best capability for using this real-time evapotranspiration information.

Soil moisture depletion can also be measured to determine when to irrigate and how much water to apply, but measurement devices such as tensiometers and resistance blocks indicate only when, not how much, to irrigate. When used in a full coverage system with full root zone exploration, however, a properly calibrated neutron probe indicates both when to irrigate and how much water to apply. Finding a representative location for placing soil moisture measurement devices can be difficult in sites irrigated with a micro-irrigation system because of the system's localized water delivery.

A good practice is to use evapotranspiration information and soil moisture measurement devices together.

Once it is decided how much irrigation water should be applied (gross irrigation), it must be determined how long the micro-volume system should operate to deliver the appropriate amount of water. The reader is referred to the chapter, *"How Much Water is Being Applied,"* on *page 71* of this handbook for guidance in determining the application rate of a micro-irrigation system.

Keeping the water application uniform is equally important to irrigating efficiently. If the system delivers water at different rates to various parts of the field, irrigation times are often based on the sections of the field receiving the least amount of water, causing some parts of the field to receive more water than other parts.

Summary

In summary, to irrigate efficiently:

• Make certain the micro-irrigation is well-designed, well-maintained, and that it applies water evenly;

• Know how much water to apply at each irrigation;

• Know the application rate of the irrigation system in inches per hour;

• Operate the system for the correct length of time.

Applying Water Uniformly

By Larry Schwankl, UC Irrigation Specialist

The more uniformly irrigation water is applied, the more efficient the irrigation. When water is applied unevenly, some parts of the field may have to be over-irrigated so that other areas receive enough water. If every emitter discharged water at the same rate, uniformity would be 100 percent.

Irrigation uniformity is most commonly measured as emission uniformity (EU), which is defined as:

$$\text{Emission Uniformity (EU-\%)} = \frac{\text{Minimum Depth of Applied Water}}{\text{Average Depth of Applied Water}} \times 100 \qquad (1)$$

where the minimum depth of applied water is often figured as the average discharge of the lowest 25% of all emitters measured. The average depth of applied water is then calculated as the average discharge of all the emitters measured. (Emission uniformity is sometimes also referred to as distribution uniformity (DU).)

A well-designed and well-maintained micro-irrigation system for orchards should have an emission uniformity of 85 to 90 percent.

Non-uniform water distribution in micro-irrigation systems stems from three main causes:

Causes of Non-Uniformity

• Because of manufacturing differences, even new emitters may vary in output. This manufacturing variation is defined by the *coefficient of variation*, or *CV*, which is the standard deviation ÷ mean value of a sampling of emitters. The manufacturer should provide this CV rating for a given system. A CV of 0.05 or less is considered good, 0.05-0.10 is average, and 0.10-0.15 is marginal.

• Differences in operating pressure within the irrigation system will cause emitters to discharge at different rates. These system pressure differences are usually caused by elevation differences within the system or by pressure being lost to friction as the water moves through pipelines, lateral lines, and fittings.

A well maintained irrigation system—one that allows for elevation differences and is designed with pipelines of the correct diameter and lateral lines of the correct diameter and length—can minimize pressure variations. In particularly difficult design situations, pressure-compensating emitters (which maintain a constant discharge even when pressure varies) may be appropriate to ensure good uniformity.

In older systems, emitters can become clogged with particulate matter (sands and silts), chemical precipitates (calcium carbonate, and iron), and biological matter (algae and bacterial slimes), which can affect emitter discharge and cause non-uniformity.

Good filtration systems and chemical treatment, where necessary, can prevent clogging and rehabilitate a clogged drip irrigation system. Since repairing a badly clogged system is difficult, a good routine maintenance program is essential.

Emission uniformity in the field should be measured periodically. The first measurement, which should occur shortly after the system is installed, will (1) reveal the quality of the irrigation system design and installation, and (2) provide a baseline by which future uniformity measurements can be compared. Future evaluations showing a decrease in emission uniformity provide warning of problems developing.

Putting a System Together

Designing a System
By Larry Schwankl, UC Irrigation Specialist

When converting from furrow, border, or sprinkler irrigation to micro-irrigation, growers must consider a number of important questions. While most micro-irrigation systems are designed by a professional designer or irrigation dealer, knowing the right questions to ask will help ensure that the system design selected best meets the grower's needs. Following is a description of the main system components and operating considerations of which the grower should be aware.

Irrigation Water Supply and Quality

Two questions should be asked before the decision to convert to micro-irrigation is made:

- Is the water available frequently enough for micro-irrigation?
- Is the water quality appropriate for micro-irrigation?

Micro-irrigation requires irrigating frequently. During peak demand periods (in the middle of the summer for most crops), daily irrigations may be required. If a well is the water source, water can be obtained "on demand," but if water is delivered by an irrigation district, it may be difficult to arrange frequent deliveries. Growers who are members of an irrigation district should contact the district to discuss the possibility of "on demand" or frequent deliveries.

The water available for the micro-irrigation system should be analyzed to determine whether it is appropriate for micro-irrigation use. Certain constituents in water can cause clogging in the small orifices of drippers and micro-sprinklers. If groundwater is to be used, the water should be analyzed for pH, iron, manganese, calcium, magnesium, carbonates, and bicarbonates. _(See "Assessing Water Quality," page 97_ and _"Chemical Precipitate Clogging," page 103_ in this handbook, for chemical constituent levels that could cause clogging.) When surface water is used, sands and silts can cause particulate clogging, and algae and bacterial slimes can cause organic clogging. But even if the water to be used presents potential difficulties, steps can be taken in designing and managing the system to mitigate clogging.

The micro-irrigation system should be designed to provide for the water needs of the fully developed crop (permanent or annual) during the peak demand period, and should be designed to operate no more than about 16 -18 hours a day during peak demand periods. This feature provides some additional irrigation capacity in the event of the need to "catch up" as a result of a system breakdown or periods of unusually high plant water demand.

Choosing a Pump

As part of the design process, the flowrate and pressure requirements at the head of the system must be determined. The pump selected should deliver the flow and pressure needed at peak efficiency. If the pump is not matched to the flow and pressure requirements, it will operate inefficiently and cause extra expense. *(See "Choosing a Pump" page 35 for more detailed information.)*

Flowmeter and Pressure Gauges

A flowmeter and pressure gauges should be part of every micro-irrigation system. By revealing the system flowrate and pressure, these devices serve two important functions — first, by indicating how much water the system is applying, they provide vital information for irrigation scheduling and make it possible for irrigation to be carried out efficiently; second, they provide a guide to the health of the system. A flowrate that decreases during the season (measured at the same pressure) may signal that clogging is occurring, whereas significant increases in flowrate may indicate a leak in the system. *(See "Flowmeters," page 41.)*

Filters

Every micro-irrigation system should have a filter. If the system uses groundwater, screen or disk filters are the usual choice; if the system uses surface water, sand media filters are used. The reader is referred to *"Filtration Equipment," page 51* in this handbook, for more details. Since clogging can occur in micro-irrigation and since good filtration is important to prevent clogging, the filter choice should take into account the quality of the available water and the filter size should be appropriate for the system flowrate and water quality.

Pressure Regulation

An operating pressure at the emitters will have been selected as part of the system design. Pressure should be kept constant, since changes in pressure affect the flowrate of most micro-irrigation emitters. A pressure-regulating valve (usually adjustable) at the head of the system or pre-set pressure regulators (usually located at the head of the laterals) can be used to keep pressure uniform. Since either solution works well, the choice of which to use may be based on the designer's preference or on design or topographical considerations.

For irrigation to be uniform, the pressure throughout the irrigation system should be nearly constant so that discharge rates are the same at each of the emitters. Pressure-compensating emitters can be used to produce a nearly constant discharge rate even when pressure varies. Pressure-compensating emitters are slightly more expensive than standard emitters and may wear out sooner because they have a flexible orifice.

Mainlines and Submains

The appropriate size for the mainlines and submains should be determined during the design phase. If mains or submains are too small, pressure loss as a result of friction — often referred to as "friction loss" — will be excessive, making extra pressure necessary at the head of the system. Excessive friction losses also cause uneven pressure and discharge rates at the emitters, making irrigation non-uniform.

When mainlines and sub-mains are too large, they inflate the cost of the system without providing any advantage. The reader is referred to _("Calculating Pressure Loss in Mainlines and Submains," page 77)_ for assistance in determining pressure losses in PVC pipelines.

Laterals

Lateral lines should be the appropriate length. If the laterals are too long, too much pressure will be lost to friction and discharge rates among the emitters will be uneven, again leading to non-uniform irrigation. The appropriate length for the lateral lines depends on the flowrate and on the number of emitters along the lateral. The objective is for emitter discharge to be the same throughout the irrigation system. The reader is referred to _"Calculating Pressure Loss in Lateral Lines," page 79_ in this handbook, for help in calculating pressure losses in lateral lines due to friction.

Injection Systems

Micro-irrigation makes it possible to inject fertilizers and other chemicals through the irrigation system. This not only simplifies delivery, but also targets the fertilizer to the root system where it will do the most good.

Several injection systems are available, including venturi injectors and injection pumps. The injection systems vary in cost and delivery accuracy. The best system to use depends on what chemicals are to be injected and on the degree of precision required. The reader is referred to _"Injection Devices," page 57_ in this handbook, for more information.

Distribution Uniformity

Uniform irrigation requires that emitter discharge rates be the same throughout the irrigation system to eliminate the need for some sections of the field to be over-irrigated in order that other sections receive enough water. The emission uniformity of drip irrigation systems can reach 80-90% in tree and vine crop systems.

The emission uniformity should be measured just after the new micro-irrigation system is installed. This practice not only tests the quality of design and installation, but also provides a baseline measure for comparing future distribution uniformity measurements. A yearly evaluation of uniformity will reveal whether the system is continuing to work well. Decreases in uniformity may indicate clogging or other problems.

Maintenance

A micro-irrigation system must be maintained properly in order to continue applying water uniformly. The designer/dealer should provide information on the recommended frequency and method of backwashing or cleaning filters, on the need to add chemicals to prevent chemical precipitate or organic clogging problems, and on how often the mains, sub-mains, and laterals should be flushed.

The irrigation system should be inspected while it is running to detect any clogged drippers or micro-sprinklers needing replacement. This inspection will also catch any leaks caused by animals chewing on the lateral lines or by traffic in the field.

To irrigate efficiently, a micro-irrigation system must not only be designed to apply water uniformly, but must also be operated properly — which requires the operator to know how much water to apply (usually in inches per day) and the application rate of the system (in inches per hour). Irrigation scheduling — determining the irrigation timing and the amount of water to be applied — can be accomplished through a variety of methods, including measuring the soil moisture, measuring plants, or determining evapotranspiration. Information on crop water use is available in many regions of the state from newspaper and radio sources and through telephone dial-in systems.

The system designer should provide the application rate (in inches/hour), but the operator can calculate this information with little difficulty. Some confusion may result from the fact that discharge rates from drippers or micro-sprinklers are usually given in gallons per hour. *"How Much Water is Being Applied?" page 71* in this handbook, can provide help in determining system application rates.

Many tree crops and vines have been successfully converted from a full coverage irrigation system (flood or sprinkler) to a partial coverage irrigation system (drip or micro-sprinkler), but the following management strategies will help ensure that the crop does not suffer as it adapts to the localized, frequent deliveries of micro-irrigation.

• Make the conversion early in the spring and begin watering with the micro-irrigation system somewhat earlier than normal. This allows the crop to take advantage of winter rainfall stored in the soil as its root development adapts to using the water delivered by the micro-irrigation system.

• If possible, apply one or two seasonal irrigations with the old, full-coverage irrigation system.

• During the first season, over-irrigate slightly with the micro-irrigation system to help ensure an adequate wetted soil volume from which the crop can draw water.

What is an Appropriate Lateral Length?
By Blaine Hanson, UC Irrigation and Drainage Specialist

Irrigating efficiently by applying water as evenly as possible over the field is achieved in micro-irrigation by minimizing differences in emitter discharge rates. Pressure variation throughout the micro-irrigation system can affect these emitter discharge rates.

The Effect of Friction on Pressure

Friction from the water flowing through the laterals (or mainlines and submains) causes pressure to decline. The diameter of the tubing, the roughness of the tubing material, the flowrate in the tubing, and the length of the laterals all determine how much friction is generated in the system and therefore the extent to which pressure varies as water flows through the system components.

Figure 1 illustrates the effect of friction on pressure along a lateral where the slope is 0% (middle set of bars). Pressure is highest at the inlet and decreases rapidly along the upper part of the lateral, then more slowly along the lower part of the lateral. The rapid decrease results from the high flowrates in the drip tubing along the upper part of the lateral, but as water is discharged through the emitters, the flowrate along the lateral gradually drops off, causing the pressure loss from friction to also decline. Along the lower part of the lateral, the lower flowrates result in only slight pressure changes.

Effect of Elevation Change on Pressure

Figure 1 also shows the effect of elevation change on pressure — in this case 1% uphill and downhill slopes. Where water is flowing uphill, pressure decreases dramatically. Each one-foot increase in elevation causes a pressure loss of 0.43 psi. When the water flow is uphill, friction and elevation act together to reduce pressure along the lateral, but when the water flow is level (0% slope), friction is the only force acting to reduce pressure.

When the water flow is downhill, gravity helps compensate for the pressure loss caused by friction. Each one-foot decrease in elevation causes a pressure increase of 0.43 psi. *Figure 1* illustrates the pressure decrease along the upper part of the lateral on a 1% downhill slope — resulting from the large amount of friction caused by the high flowrates in the lateral. In the lower part of the lateral, however, pressure becomes greater and greater with distance along the lateral. Here the low lateral flowrate results in only slight pressure loss from friction — less than the pressure gains due to the effects of gravity — and pressure instead increases near the end of the lateral.

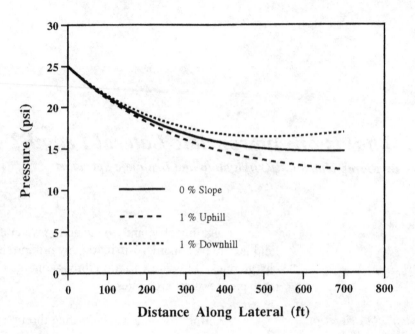

Figure 1. Effect of friction on pressure along a 0% slope, a 1% downhill slope, and a 1% uphill slope lateral.

The Relationship Between Tubing Diameter and Pressure

The larger the diameter of the drip tubing, the less pressure lost to friction, assuming equal flowrates. Larger diameter drip tubing therefore allows for longer lateral lengths without adverse effect on emitter discharge uniformity. Doubling the tubing diameter reduces the friction loss by about 29 times. Thus, friction losses are highly sensitive to tubing diameter.

Flowrate and Pressure Loss

The higher the flowrate in the lateral, the greater the pressure loss from friction. Doubling the flowrate increases the friction loss by 3.6 times. The lateral flowrate depends on lateral length, emitter spacing and emitter design. The smaller the emitter spacing, the higher the flowrate for a given emitter lateral. For a given lateral length, smaller spacings result in more emitters. The longer the lateral length, the higher the flowrate.

Emitter Spacing

Emitter spacing can also affect the flowrate — and therefore the pressure — along a drip lateral. With a given set of emitter characteristics, the closer the emitter spacing, the higher the flowrate and the greater the pressure losses along the lateral. The closer the emitter spacing, the shorter the lateral length necessary to maintain good uniformity.

How Pressure Affects Discharge Rates

The effect of pressure on the emitter discharge rate is defined by the emitter discharge exponent. The smaller the exponent, the less effect pressure variation has on discharge rate. *Figures 2* and *3* illustrate discharge rates for emitter discharge exponents of 0.575 and 0.1, respectively. Pressure affects the discharge rate of the emitter with the large exponent much more than the discharge rate of the emitter with the small exponent.

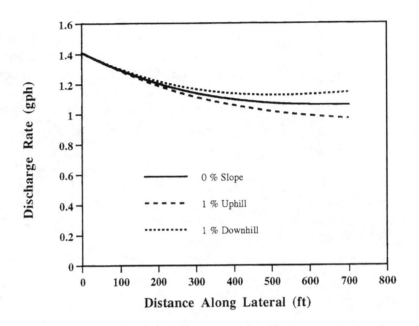

Figure 2. Drip emitter discharge rates for a lateral line with an emitter discharge exponent of 0.575.

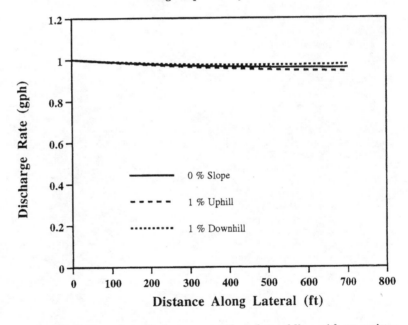

Figure 3. Drip emitter discharge rates for a lateral line with an emitter discharge exponent of 0.1.

Coping With Pressure Variation

Options available for reducing pressure variation or its effect on emitter discharge rate include:

• Design the irrigation system such that pressure variation due to friction and elevation differences throughout the system is small by properly sizing both diameters and lengths of pipelines and drip tubing.

• Install pressure regulating valves throughout the system. One approach is to split the drip system into subunits with a pressure regulating valve at the inlet of each subunit. See *"Valves and Regulators"*.

• Use emitters that are less sensitive to pressure variation, as indicated by the emitter discharge exponent.

Emission Devices
By Larry Schwankl, UC Irrigation Specialist

Micro-irrigation systems do not wet the entire the soil surface, but rather make small, frequent discharges of water to the soil through emission devices or _emitters_, making it possible to control where the water is placed. Emitters in micro-irrigation systems may be drip emitters, line sources, or micro-sprinklers.

Drip Emitters

Drip emitters provide a localized water point source, which makes them appropriate for tree and vine crops. They are also well suited for use in landscapes with stand-alone trees, shrubs, and other plants. To meet both the plant water requirements and the soil water intake, and to achieve a sizeable wetted soil volume, multiple emitters per plant may be required.

Drip emitters are configured in various ways so as to dissipate pressure. In orifice emitters, a small passageway or opening dissipates energy. In long-path emitters, long, circuitous passageways reduce pressure through wall friction. In turbulent flow emitters, tortuous paths reduce the pressure. Turbulent flow emitters _(Figure 1)_ are the most commonly used, since this configuration allows for larger passageways and is therefore less susceptible to clogging.

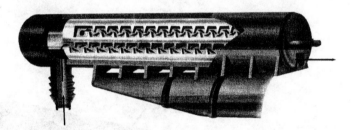

Figure 1. Turbulent flow drip emitter.

Drip emitter discharge rates range from one-half to four gallons per hour (gph), with one-half-, one-, and two-gph emitters being the most common. A one-gph emitter discharges one gallon per hour at a specific pressure. Changing the pressure at which the system is operated can change the emitter flowrate, with discharge increasing as the pressure increases. _(Figure 2)_. Pressure-compensating emitters, however, change discharge rates only slightly when operating pressures are changed _(Figure 2)_.

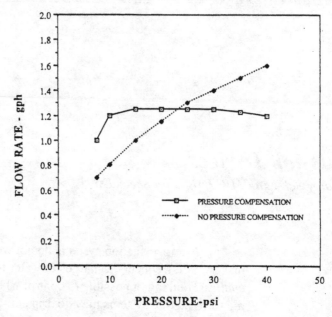

Figure 2. Emitter discharge/pressure relationship for pressure-compensating drip emitters and non-pressure-compensating emitters.

The orifice size of the passageways within a drip emitter are quite small (0.025 - 0.060 inches), and are therefore susceptible to clogging. Good filtration and chlorine and/or acid injection may be required to minimize clogging problems.

Line source emitters consist of drip tapes *(Figure 3)* or tubings with emission points at regular intervals. Drip tapes are made up of flexible tapes with walls ranging from about 4 to 25 mils (a mil = 0.001 inch) in thickness. Energy dissipates as water passes along tortuous paths. This drip tape known as turbulent flow tape, allows the use of larger passageways and is thus less susceptible to clogging.

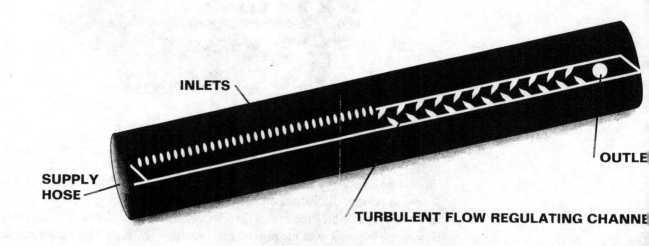

Figure 3. Drip tape.

Drip tubings have emitters built into the polyethylene tubing and operate like drip emitters. Some drip tubings are pressure-compensating.

Line source tapes and tubings are well-suited for row crops and are often used with strawberries and vegetable crops and, to a limited extent, on cotton and processing tomatoes. Subsurface drip tape or tubing is sometimes used on trees and vines. While most drip emitters are not meant to be used below ground, many tape and tubing products are designed for subsurface use. (The discharge rate of drip tape is usually given in gallons per minute per 100 feet (gpm/100 ft.).)

Micro-Sprinklers

Micro-sprinklers have a discharge rate of 4-30 gph — higher than drip emitters — and wet a larger surface area (8 to 30 feet in diameter). They are used primarily on widely spaced tree crops since these features make them well-suited to meeting the substantial water demands of a mature orchard. Micro-sprinklers are often also the choice where a large wetted area is needed in soils in which water does not move well laterally or where water infiltration is poor.

Some micro-sprinklers have moving parts (spinners), whereas others contain no moving parts, but consist instead of an orifice and fixed plate that directs the water into "fingers" of stream. *(See Figure 4.)* Various plate shapes, resulting in different wetted patterns, are available.

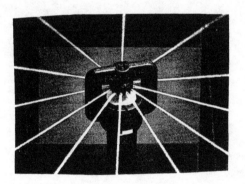

Figure 4. Micro-sprinkler.

The flow passageways of micro-sprinklers are of the same magnitude or slightly larger (0.035 - 0.080 inches) than drip emitters, with the higher discharge micro-sprinklers having larger flow passages, which makes them less susceptible to clogging. The discharge velocity from the micro-sprinkler is greater than that of a drip emitter, further minimizing the clogging hazard. Filtration and water treatment requirements for micro-sprinklers are similar to those of drip emitters.

Selecting Drip Emitters and Micro-sprinklers
By Blaine Hanson, UC Irrigation and Drainage Specialist

Factors to Consider

Drip emitters and micro-sprinklers should be chosen to obtain the maximum possible emission uniformity. The factors to be considered in making the selection are:

- the manufacturing *coefficient of variation* (CV) — that is, differences in output among emitters resulting from the manufacturing process

- the relative sensitivity of the emitter discharge rate to variations in water pressure

- the sensitivity of the emitters to clogging

- cost

Coefficient of Variation(CV)

The CV for any given emitter — which the manufacturer should provide — is calculated by measuring the flowrate of a sampling of emitters at a constant pressure. A CV of 0.05 or less is considered good, 0.05-0.10 is acceptable, and more than 0.1 is marginal.

Emitter Discharge Exponent

The emitter discharge rate is affected by changes in water pressure, with some emitters more able than others to compensate for pressure variation. This relative pressure-compensating ability is described by the *emitter discharge exponent*. The smaller the exponent, the less sensitive the discharge rate of a given emitter to pressure changes. An exponent of zero means the emitter can compensate completely for any pressure variation; an exponent of one means the emitter has no capacity to compensate for pressure variation; and an exponent of 0.5 means the emitter can partially compensate for pressure variation. If the exponent is 0.5, a 20 percent change in pressure will cause a 10 percent change in emitter flowrate.

Clogging Potential

The relative sensitivity of a particular emitter to clogging is influenced by the dimensions and configurations of the flow passages and any consequent turbulence in the water flowing through the emitter. Generally, large flow passages suggest less potential for clogging, but actual clogging tendency should be determined from field experience and from manufacturers' recommendations.

**CV s and Discharge
Exponents of Four
Emitter Types**

Table 1 shows the coefficients of variation and discharge exponents of four different emitter types.

Table 1. Coefficients of variation and discharge exponents of four emitter types*.

Emitter type	Coefficient of variation	Emitter discharge exponent
1	0.04	0.48
2	0.29	0.51
3	0.05	0
4	0.22	1.14

* Derived from test data taken from various sources

Emitters 1 and 3 have coefficients of variation (CVs) of 0.04 and 0.05, which are considered excellent, but the CVs of Emitters 2 and 4 are unsatisfactory.

The emitter exponents of the four emitters range from 0 to more than 1. Emitter 3, with an emitter exponent of 0, is completely pressure-compensating, meaning that the emitter discharge rate is not affected by variations in pressure. (Pressure compensation will only occur, however, if the pressure exceeds a minimum pressure — usually 8 to 10 psi. If the pressure falls below that level, the emitter will not completely compensate for pressure changes.)

Emitter 4, with an emitter exponent of more than 1, has no pressure-compensating capability. Emitters 1 and 2, with emitter exponents of about 0.5, are partially pressure-compensating, and are typical of many turbulent-flow emitters. In Emitters 1 and 2, a 20 percent change in pressure will cause about a 10 percent change in discharge rate.

Many drip emitters have CVs of less than 0.05. Turbulent-flow emitters have discharge exponents of about 0.5. Other drip emitters are highly pressure-compensating with exponents ranging from nearly zero (completely pressure-compensating) to about 0.25.

Micro-sprinklers

Micro-sprinklers generally have CVs of less than 0.05, although a few have CVs ranging from 0.05 to 0.1. Discharge exponents for micro-sprinklers are about 0.5, while micro-sprinklers with complete pressure-compensating capability, have exponents ranging from nearly 0 to about 0.2.

Reference

Drip emitters and micro-sprinklers. 1993. Irrigation equipment performance report, CATI Publication No. 93-1002. Center for Irrigation Technology, California State University, Fresno.

Pressure-Compensating Emitters

By Larry Schwankl, UC Irrigation Specialist, and
Terry Prichard, UC Water Management Specialist

Pressure-compensating emitters, both drip emitters and micro-sprinklers, deliver a constant discharge of water even when operating pressure varies (see *Figure 1*). Built into the emitter is a flexible orifice that opens or closes depending on the pressure, thus regulating the flow through the emitter. Below a minimum water pressure the emitter will no longer compensate for pressure changes.

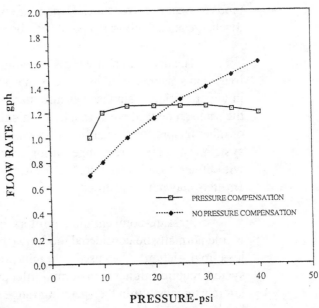

Figure 1. Emitter discharge/pressure relationship for pressure-compensating drip emitters and non-pressure-compensating emitters.

Where Should They Be Used?

Most micro-irrigation systems use one or more pressure-regulating valves to maintain the desired pressure within the irrigation system. These valve systems — which may consist of a single pressure-regulating valve at the head of the system or pressure regulators installed at the head of each lateral — do not guarantee that the pressure (and thus the flow) will be the same at all emission points. Pressure differences along a lateral resulting from changes in elevation or from pressure loss because of friction will still cause variation in emitter discharge rates, but using pressure-compensating emitters will keep emitter discharge rates uniform even if the pressure does vary. Pressure-compensating emitters are suitable for use in irrigation systems where elevation differences are significant or where lateral lengths are long.

**Advantages and
Disadvantages**

The main advantage of pressure-compensating emitters is this capacity to maintain a uniform water discharge despite pressure differences. Without pressure-compensating emitters, the emitters at the bottom (downhill) section of a lateral line running steeply downward would discharge at a significantly greater rate (because of increased pressure) than emitters on the uphill portion of the lateral.

Some manufacturers claim that pressure-compensating emitters also have less tendency to clog than standard emitters because the flexible orifice helps purge particulate matter.

Pressure-compensating emitters are more expensive than standard emitters, however, and have a shorter service life because the flexible orifice tends to wear out over time. Depending on the circumstances, ensuring uniform discharge at each emission point may be well worth the additional cost.

The decision to use pressure-compensating emitters in a micro-irrigation irrigation system should be made on a case-by-case basis. An important objective in any irrigation design is to ensure that all portions of the irrigated area receive the same amount of water (application uniformity). If topographical conditions or other factors cause significant pressure differences within the irrigation system, using pressure-compensating emitters may be desirable. For systems with little pressure variation, the additional cost of pressure-compensating emitters may not be justified.

Pressure-compensating emitters may be used with lateral lengths that would normally be considered too long (therefore causing too great a pressure loss from friction) when used with non-pressure-compensating emitters. Such a system requires higher than normal inlet pressures to ensure that pressures in the laterals remain within the operating range of the pressure-compensating emitters. Under certain field conditions, following this design strategy may eliminate the need for additional submains. Economics plays a major role in evaluating the feasibility of such a design.

Choosing a Pump
By Blaine Hanson, UC Irrigation and Drainage Specialist

The two kinds of pumps most often used to power irrigation systems are deep-well turbines and centrifugal pumps.

Deep-well Turbines

Deep-well turbines are installed inside the well casing and consist of a pump casing, impellers (located inside the casing), and shaft, which connects the impeller to the power source. Often deep-well turbine pumps include several impeller/casing combinations (each referred to as a _stage)_ to provide enough pump output.

Centrifugal Pumps

Centrifugal pumps usually consist of one impeller, a pump casing (called a _volute),_ and a shaft. The shaft and the impeller rotate, creating centrifugal forces in the water inside the impeller. These forces cause water to flow to the outer edge of the impeller and into the impeller eye or center. The amount of pressure developed by the impeller depends on the impeller diameter and rpm, while the impeller flowrate is determined by the impeller width and diameter.

Pump Performance Curves

The performance of a given pump is defined by _pump performance curves,_ which describe the relationship between the total head developed by the pump at a particular capacity (the energy imparted to the water), the pump efficiency at that capacity, and the brake horsepower requirements of the pump. These curves make it possible to identify the most efficient pump for a new system by showing which pump will provide the total head and capacity the system needs. Pump performance curves can also be used to evaluate the performance of existing pumps.

Figure 1 illustrates a total head/capacity performance curve of a pump. The total head is measured in the field as the sum of the pumping lift (the distance from the soil surface down to the pumping water level) and the discharge pressure head (discharge pressure in pounds per square inch multiplied by 2.31). The figure shows that as the capacity increases, total head decreases, which is the reason pump capacity decreases as groundwater levels decline. Maximum total head occurs at shutoff (no capacity).

Total head capacity curves take a variety of shapes. Steep curves denote little change in pump capacity as total head changes, while flat curves denote significant changes in pump capacity with changes in total head.

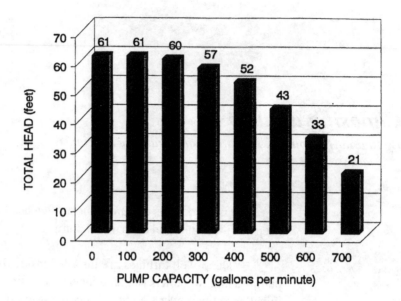

Figure 1. Total head/capacity performance curve.

Figure 2, which illustrates the pump efficiency/capacity of the same pump, shows that as capacity increases, pump efficiency also increases, until a maximum efficiency is reached. After that point efficiency decreases as capacity increases. The significance of this phenomenon is that the horsepower demand of the pump needed to provide the desired output is minimized. (Note that efficiency here refers only to *pump* efficiency.) The efficiency shown in pumping plant test data is the overall efficiency — the pump and motor efficiency combined.

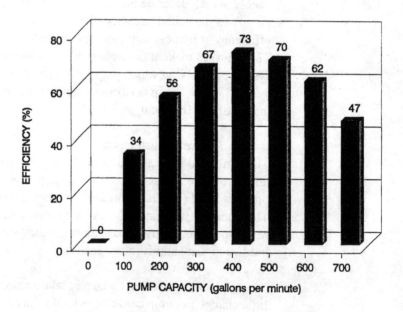

Figure 2. Pump efficiency/capacity performance curve.

The brake horsepower/capacity curve shown in _Figure 3,_ illustrates that as capacity increases, brake horsepower also increases. In some deep-well turbine impellers, the horsepower increases to a maximum point, while in some centrifugal pump impellers, horsepower continues to increase as capacity increases.

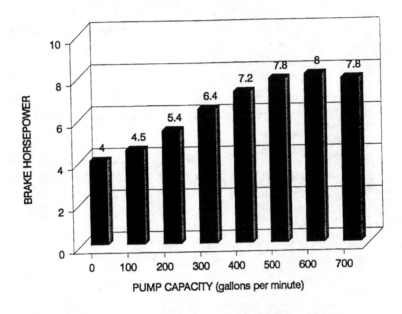

Figure 3. Brake horsepower/capacity curve.

Finding the Most Efficient Pump

The first step in identifying which pump is best suited to a particular irrigation system is measuring the pumping lift and determining the pump capacity and discharge pressure needed. The necessary capacity is determined by the crop's water requirements, the efficiency of the irrigation system, and the size of the acreage to be irrigated. The discharge pressure needed depends on the pressure requirements of the irrigation system, which in turn depends on the desired operating pressure of the emitters, on how much pressure is lost to friction as water moves through the system, and on elevation differences in the system. The pumping lift will have to be estimated from pumping tests on the existing well or on nearby wells. The pumping lift and the discharge pressure are added together to calculate total head.

Once the total head and pump capacity have been determined, catalogs showing pump performance curves can be consulted to find a pump that will provide the desired capacity and total head at close to maximum efficiency, which will result in a minimum horsepower demand for that capacity and total head. The brake horsepower/capacity curve is used to determine the motor or engine size.

Example

Table 1 illustrates the importance of choosing an efficient pump. Each of the three pumps *A, B,* and *C* provides the same output (940 gallons per minute, 228 feet of total head), but the pump efficiency ranges from 69% to 84%. Brake horsepower ranges from 64 (for the 84% efficiency) to 73.5 (69% efficiency). At 1500 annual operating hours and an electric energy rate of $0.10 per kilowatt-hour, the annual energy cost ranges from $7162 (84% efficiency) to $8225 (69% efficiency).

Table 1. Effect of pump efficiency on annual energy cost.

	Pump A	Pump B	Pump C
pump efficiency (%)	84	69	76
brake horsepower	64	73.5	69
annual energy cost ($)	7162	8225	7721

Another water management handbook in this series, *Irrigation Pumping Plants* (Publication No. 93-04), provides more detailed information about irrigation pumps.

Valves and Regulators

By Blaine Hanson, UC Irrigation and Drainage Specialist, and
Larry Schwankl, UC Irrigation Specialist

Valves and regulators to control pressure and water flow are essential in a drip irrigation system. The following are recommended.

Check Valves

Check valves, usually installed at the pump discharge, prevent water from flowing backward into the well. Check valves are also an important component of injection systems to prevent reversal of flow which might cause a chemical storage tank to fill with irrigation water and overflow. They also prevent low-head drainage of chemicals into the irrigation system. These may be either of two types: swing check valves or double disc valves.

Pressure-Regulating Valves

Pressure-regulating valves, often referred to as pressure-reducing valves, dissipate upstream pressure to help keep downstream pressure constant, but do not otherwise control the water flow. Their placement in the irrigation system is a design decision often based on considerations such as topography. Design option include: (1) a single, large, adjustable, pressure-regulating valve may be installed at the head of the system; (2) multiple, smaller, adjustable pressure-regulating valves strategically placed in the submains; or (3) a small, pre-set, pressure regulators placed at the head of each lateral.

Some irrigation designs incorporate a partially-closed value(s), such as a gate valve, to reduce operating pressure. While this reduces pressure, due to frictional head loss, it has no capability of adjusting to changes in system flow or pressure. In addition, if numerous partially closed valves are used, they all must be re-adjusted whenever a change is made in one valve.

Pressure-Relief Valves

Pressure-relief valves, which are usually spring-loaded to open when pressure exceeds a designated level, protect against pressure surges that might damage pipelines or filter housings.

Noncontinuous Air Vents

Noncontinuous air vents allow air to escape from the pipeline while it is filling. These vents should be placed just downstream from the pump, at high points along the pipeline, or at places where air could be trapped by water while the pipeline is filling. Once the air escapes, the vent closes to prevent water from flowing out of the pipeline.

Continuous Air Vents

Continuous air vents also function to allow air to escape from the pipeline during filling, but these continue releasing air from the pipeline while the water is under pressure. These vents may be needed where air has become trapped (entrained) in the water as a result of water cascading down into the well from a point above the water level.

Vacuum-relief Valves or Vents

Vacuum-relief valves or vents allow air to enter the pipeline after the pump is shut off and the pipeline is draining. This prevents a vacuum from forming in the pipeline.

Flow Control Devices

Flow control devices keep water flowing into the pipeline at a constant rate, despite changes in pressure. These devices usually consist of flexible orifices or membranes that change the flow passage dimensions as the pressure changes. Once the flow control device is in place the flowrate cannot be changed by varying the upstream or downstream pressure. The best flow control device to use depends on the flowrate desired.

Flowmeters

By Blaine Hanson, UC Irrigation and Drainage Specialist, and
Larry Schwankl, UC Irrigation Specialist

Flowmeters measure the volume of water moving through a full-flowing closed pipe and as such are one of the key components of a drip irrigation system. They are essential for managing irrigation efficiently and for monitoring the performance of the irrigation system. Managing irrigation efficiently requires: (1) knowing how much water the crop has used since the last irrigation (irrigation scheduling); and (2) operating the irrigation system to apply only the amount of water desired. A flowmeter gives the manager the information needed to apply only the amount of water required.

Monitoring the performance of a micro-irrigation system makes it possible to identify changes in flowrate during the season (measured at the same pressure), which may indicate problems such as clogging of emitters or filters, leaks in the system, or problems with the pump or well.

Propeller Flowmeters

Propeller flowmeters, consisting of a propeller linked by a cable or shafts and gears to a flow indicator and inserted into a pipeline, are frequently used to measure flowrates in pipelines. The indicators can report either the flowrate or the total flow volume, or both.

Propeller flowmeters can be installed in several different ways: by being inserted into a short section of pipe, which is then either coupled *(Figure 1a)*; bolted *(Figure 1b)* into the pipeline; clamped, strapped, or welded onto the pipeline as a saddle-type meter *(Figure 1c)*; or inserted into the pipe discharge *(Figure 1d)* to measure flow from gates installed in canals and ditches.

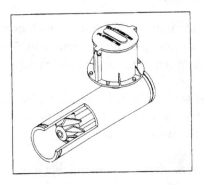

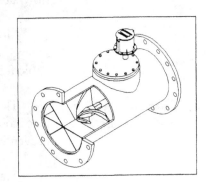

Figure 1a. Welded propeller flowmeter.

Figure 1b. Bolted propeller flowmeter.

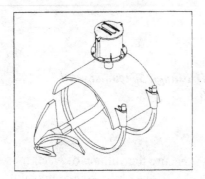

Figure 1c. Saddle-type propeller
flowmeter.

Figure 1d. Insertion-type
propeller flowmeter.

Illustrations courtesy of Ketema/McCrometer Division, Hemet, CA 92545

Propeller meters can also be installed as in-line meters in a short section of portable pipe *(Figure 1a)*. Couplings are welded onto each end of the flowmeter pipe section to connect sections of the portable pipe. One flowmeter of a given size can be used for several sizes of portable pipe if a sufficiently long straight section of pipe of the same diameter as the flowmeter section is installed immediately upstream from the flowmeter.

Since clamp-on and strap-on saddle meters can be moved from location to location, one flowmeter can be used to measure flowrates for several pumping plants. The same pipe diameter must be used at each site, however. Dummy saddles can be installed at locations other than the one at which the meter is installed.

Selecting Propeller Flowmeters

Propeller meters must be matched to the correct pipe size — since the gear mechanism connecting the propeller to the indicator is based on the pipe inside diameter — and to the desired flowrate. *Table 1* gives maximum and minimum flowrates for various pipe diameters.

Most flowrate indicators report in gallons per minute or in cubic feet per second, while total flow indicators ("totalizers") report in gallons, acre-feet, or cubic feet. Some indicators report in metric units.

Installation and Operation

The flowmeter should be installed at a location of minimal water turbulence, since too much turbulence will cause the flowrate indicator to oscillate wildly, preventing reliable measurement.

Manufacturers often recommend that a section of straight pipe, eight to ten pipe diameters long, be placed immediately upstream from the flowmeter, but field experience has shown that a longer pipe length may be required in locations where jetting may occur because of a partially closed valve. One manufacturer maintains, however, that reliable measurements can be made with even relatively short sections of pipe if straightening vanes are used.

Table 1. Flowmeter maximum and minimum flowrates for various pipe diameters.
(Courtesy of McCrometer Flowmeters)

Meter and nominal pipe size	4	6	8	10	12	14
Maximum flow rate	600	1200	1500	1800	2500	3000
Minimum flow rate	50	90	108	125	150	250

Swirling

A centrifugal sand separator or a series of elbows in the pipeline may cause a swirl or rotation to develop in the flowing water. The swirls may still be present even one hundred pipe diameters downstream. The remedy is to place straightening vanes in the pipeline just in front of the flowmeter. One manufacturer recommends that a six-vane straightener be used.

A relatively stable propeller meter rate indicator means that turbulence is minimal, but wide variation in indicator readings signals turbulence in the pipeline. If the rate indicator shows erratic, violent behavior, air or gas may be present in the water.

Attaining Full Flow

Propeller flowmeters will operate properly only if the pipe is flowing full. If the pipeline is only partially full, the flowrate measurement will not be accurate. In pressurized irrigation systems, flow will usually be full at the pump discharge, but in pumps with an open discharge, as into an irrigation ditch, pipe flow may not be full. The problem can be remedied by creating a slight rise in the discharge pipe; by installing a gooseneck at the pipe discharge; by installing an elbow (discharge end pointing upward at the pipe discharge); or by installing a valve downstream from the meter.

Accuracy

Under ideal conditions, a propeller flowmeter operated within its recommended range can be accurate to within ± 2 percent, but if the flowrate is too slow, accuracy will be less.

Pressure/Head Loss

Inserting a propeller into the water flow can cause friction, resulting in pressure or head losses in the pipeline. The amount of pressure lost depends on the velocity or flowrate and on the pipe diameter. The higher the flowrate, the more pressure lost because of the flowmeter, but the larger the pipe diameter, the less pressure lost.

As *Figure 2* illustrates, these pressure losses are generally small. With a ten-inch flowmeter, the pressure loss is less than 0.1 psi for flowrates less than 2000 gpm.

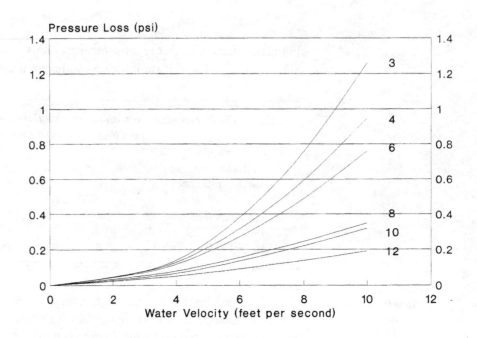

Figure 2. Pressure loss caused by propeller flowmeters.
(Figure developed from data supplied by Water Specialties Corporation, Porterville, CA 93257.)

Magnetic Flowmeters

The recently introduced magnetic flowmeter has the advantage of not causing an obstruction in the pipe. This feature eliminates the problem of possible entanglement from debris in the water as well as any pressure loss across the device. Magnetic flowmeters also require less maintenance than propeller meters, have long-term accuracy, and can be installed only five pipe diameters of straight pipe upstream from the meter, but have the disadvantages of a higher initial cost and the need for an external power supply.

Ultrasonic Flowmeters

Ultrasonic flowmeters measure flow velocity (and thus flowrate) by directing ultrasonic pulses diagonally across the pipe both upstream and downstream. The difference in time required for the signal to travel through the moving water is measured and converted to flow velocity. Ultrasonic flowmeters have accuracy comparable to that of propeller meters and, since they have no moving parts, require little maintenance. Because all attachments are external, these meters can be moved easily to different locations. Since the ultrasonic meters work by bouncing the ultrasonic pulses off particles in the water, irrigation waters which don't have enough suspended particles in it may not be appropriate for ultrasonic meter use. It has been found that high quality well waters may fall into this category. Ultrasonic flowmeters generally cost more than other types of meters.

Turbine Flowmeters

Turbine flowmeters operate on the principle of a rotor assembly, turning at a rate proportional to the flowrate in the pipelines. The rotor is suspended near a magnetic pickup, which records a pulse on its readout unit as each rotor blade passes. Turbine flowmeters have an accuracy comparable to that of propeller flowmeters (within a few percent) under the correct flow conditions and require a ten-pipe diameter length of straight pipe upstream and a six-pipe diameter length downstream. The turbine flowmeter is more sensitive to non-uniform flow conditions, such as exist downstream of an elbow or constriction, than is a propeller flow meter. Some turbine flowmeters can be installed in a range of pipe diameters, which provides flexibility in their use. The applicability and cost are comparable to that of a propeller meter.

Venturi Flowmeters

Venturi flowmeters consist of a section of pipe with a restriction in a specific shape, across which pressure change is measured. The magnitude of this pressure change depends on the flowrate through the device. Venturi flowmeters offer the advantages of unobstructed water flow, no moving parts (meaning low maintenance requirements), little pressure loss across the device, and good accuracy. Venturi flowmeters cost slightly more than propeller meters and the meter readout is less convenient than that of some other flowmeters.

Converting Flowmeter Readings

The readout from flowmeters can be in instantaneous flowrate in gallons per minute (gpm) or cubic feet per second (cfs), or in total flow in gallons, cubic feet, or acre-feet. For irrigation scheduling purposes, however, crop water use is given in inches or in inches per day. The following formulas can be used to convert flowmeter readout to inches per hour.

____ gpm ÷ area irrigated (acres) x 0.0022 = ____ in/hr (see also *Table 2*).

____ cfs ÷ area irrigated (acres) x 0.992 = ____ in/hr (see also *Table 3*)

____ gallons ÷ time period over which measured (min) ÷ area irrigated (acres)
    x 0.0022 = ____ in/hr

____ cubic feet ÷ time period over which measured (minutes) ÷ area irrigated
    (acres) x 0.0165 = ____ in/hr

____ acre-feet ÷ time period (minutes) ÷ area irrigated (acres) x 720 = ____ in/hr

Using the Flowrate to Find Out How Much Water is Being Applied During an Irrigation Set

From the flowrate measurement, the amount of water applied during an irrigation set can be calculated in inches using the following equation:

$$D = (Q \times T) \div (449 \times A) \tag{1}$$

where   D = inches of applied water
          T = actual hours required to irrigate the field
          A = acres irrigated
          Q = flowrate in gallons per minute

| Example |

Calculate the inches of water applied if a flowrate of 300 gallons per minute is used to irrigate 40 acres in 16 hours.

$$D = (300 \text{ gpm} \times 16 \text{ hours}) \div (449 \times 40 \text{ acres}) = 0.27 \text{ inches}$$

References

Ketema/McCrometer Division. *McCrometer propeller flowmeters: manual for installation, operation, and maintenance.*

Ketema/McCrometer Division. *Bulletin G1200. Basic specifications for McCrometer propeller flowmeters.*

Ketema/McCrometer Division. *Bulletin MR-300. McCrometer flowmeters.*

Noffke, Milvern H. 1991. "Achieving accurate irrigation water measurements with propeller meters." ASCE Specialty Conference, *Planning Now for Irrigation and Damage in the 21st Century,* July 18-21, 1988, Lincoln, Nebraska.

Table 2. Application rate (in/hr) for known acreage and flowmeter flowrate (gpm).

APPLICATION RATE - INCHES PER HOUR

FLOW RATE GPM	IRRIGATED AREA - ACRES								
	1	2	5	10	15	20	25	50	100
10	0.022	0.011							
15	0.033	0.016	0.007						
20	0.044	0.022	0.009	0.004					
25	0.055	0.027	0.011	0.005					
50	0.110	0.055	0.022	0.011	0.007				
75	0.165	0.082	0.033	0.016	0.011	0.008	0.007		
100	0.220	0.110	0.044	0.022	0.015	0.011	0.009		
150	0.330	0.165	0.066	0.033	0.022	0.016	0.013	0.007	
175		0.192	0.077	0.038	0.026	0.019	0.015	0.008	
200		0.220	0.088	0.044	0.029	0.022	0.018	0.009	
250		0.275	0.110	0.055	0.037	0.027	0.022	0.011	
300		0.330	0.132	0.066	0.044	0.033	0.026	0.013	0.007
350			0.154	0.077	0.051	0.038	0.031	0.015	0.008
400			0.176	0.088	0.059	0.044	0.035	0.018	0.009
500			0.220	0.110	0.073	0.055	0.044	0.022	0.011
600			0.264	0.132	0.088	0.066	0.053	0.026	0.013
700			0.308	0.154	0.103	0.077	0.062	0.031	0.015
800				0.176	0.117	0.088	0.070	0.035	0.018
900				0.198	0.132	0.099	0.079	0.040	0.020
1000				0.220	0.147	0.110	0.088	0.044	0.022
1100				0.242	0.161	0.121	0.097	0.048	0.024
1200				0.264	0.176	0.132	0.106	0.053	0.026
1300				0.286	0.191	0.143	0.114	0.057	0.029
1400				0.308	0.205	0.154	0.123	0.062	0.031
1500				0.330	0.220	0.165	0.132	0.066	0.033
2000					0.293	0.220	0.176	0.088	0.044
2500						0.275	0.220	0.110	0.055
3000						0.330	0.264	0.132	0.066
3500							0.308	0.154	0.077
4000								0.176	0.088
4500								0.198	0.099
5000								0.220	0.110

Table 3. Application rate (in/hr) for known acreage and flowmeter flowrate (cfs).

		APPLICATION RATE - INCHES PER HOUR								
		IRRIGATED AREA - ACRES								
		1	2	5	10	15	20	25	50	100
	0.01	0.010								
	0.02	0.020	0.010							
	0.03	0.030	0.015	0.006						
	0.04	0.040	0.020	0.008						
	0.05	0.050	0.025	0.010						
	0.1	0.099	0.050	0.020	0.010	0.007				
	0.2	0.198	0.099	0.040	0.020	0.013	0.010	0.008		
	0.3	0.298	0.149	0.060	0.030	0.020	0.015	0.012	0.006	
	0.4		0.198	0.079	0.040	0.026	0.020	0.016	0.008	
	0.5		0.248	0.099	0.050	0.033	0.025	0.020	0.010	
	0.75			0.149	0.074	0.050	0.037	0.030	0.015	0.007
	1			0.198	0.099	0.066	0.050	0.040	0.020	0.010
FLOW	1.25			0.248	0.124	0.083	0.062	0.050	0.025	0.012
RATE	1.5			0.298	0.149	0.099	0.074	0.060	0.030	0.015
CFS	1.75				0.174	0.116	0.087	0.069	0.035	0.017
	2				0.198	0.132	0.099	0.079	0.040	0.020
	2.25				0.223	0.149	0.112	0.089	0.045	0.022
	2.5				0.248	0.165	0.124	0.099	0.050	0.025
	2.75				0.273	0.182	0.136	0.109	0.055	0.027
	3				0.298	0.198	0.149	0.119	0.060	0.030
	3.25					0.215	0.161	0.129	0.064	0.032
	3.5					0.231	0.174	0.139	0.069	0.035
	3.75					0.248	0.186	0.149	0.074	0.037
	4					0.265	0.198	0.159	0.079	0.040
	4.25					0.281	0.211	0.169	0.084	0.042
	4.5					0.298	0.223	0.179	0.089	0.045
	4.75						0.236	0.188	0.094	0.047
	5						0.248	0.198	0.099	0.050
	5.5						0.273	0.218	0.109	0.055
	6						0.298	0.238	0.119	0.060
	6.5							0.258	0.129	0.064
	7							0.278	0.139	0.069
	7.5							0.298	0.149	0.074
	8								0.159	0.079
	8.5								0.169	0.084
	9								0.179	0.089
	9.5								0.188	0.094
	10								0.198	0.099

How Deep to Place a Subsurface System

By Larry Schwankl, UC Irrigation Specialist, and
Blaine Hanson, UC Irrigation and Drainage Specialist

Several factors must be considered in deciding how deep subsurface drip tapes or tubing should be buried, including: type of crop, relative permanance of the system, installation needs, crop germination requirements, and weed control.

Type of Crop

Because of differences in root development, traffic, germination, and weed control, subsurface systems for irrigating permanent crops are usually buried deeper than systems intended for annual crops. Usual depths for permanent crop systems are 9 to 18 inches, while those for annual crops are 2 to 9 inches.

How Permanent Will the System Be?

Subsurface drip systems intended to last for several years are usually placed deeper than systems intended to remain in place for only a season or two. If the system is to remain for several seasons, the tape/tubing is often placed at least 8 inches deep to allow for cultivating, preparing beds, and harvesting, while single-season installations – as for strawberries and vegetables – may be placed only a few inches deep.

Installation

Generally, the deeper the installation, the more difficult it is to install the tape/tubing without damage from stretching or breaking. The horsepower needed to pull the installation shank(s) through the field also increases with burial depth.

Establishing permanent crops with subsurface drip systems requires careful planning. Since newly planted trees and vines have a limited root zone, some growers use a surface drip system with emitters placed near the young plant for the first season or two and then switch to subsurface drip irrigation with emitters placed farther from the trunk after the root zone has expanded. This option increases irrigation system costs.

Controlling Weeds

In both annual and permanent crops, subsurface drip systems have the major advantage of reducing weeds by keeping much of the soil surface dry. Drip system burial depths of 9 inches or more have been shown to be effective in reducing summer weed growth. In permanent crops, using drip tubing with emitter discharge rates of about 0.5 gph results in less "surfacing" of water than using tubing with higher emitter discharge rates. Various other management techniques such as: disking following the shanking in of the tubing, allowing the disturbed soil to re-settle prior to the drip tubing operation, or flood or sprinkler irrigating to settle the disturbed soil have all been tried with varying success.

It should be noted that in row crops, germination and weed control considerations are at odds with each other, since shallower burial depths mean better seed germination but poorer weed control.

The following recommended subsurface drip irrigation system burial depths reflect state-of-the-art research and grower experience:

- single-season row crop systems: 2 to 6 inches
- multi-season row crop systems: 8 to 10 inches
- permanent vines: 8 to 12 inches
- permanent trees: 12 to 24 inches

Filtration Equipment

By Larry Schwankl, UC Irrigation Specialist,
Blaine Hanson, UC Irrigation and Drainage Specialist, and
Terry Prichard, UC Water Management Specialist

Since drip emitters have small passageways that can become clogged by inorganic materials (sands, silts, clays, and the chemical precipitates lime and iron) or by organic materials (algae, bacteria, and the slimes they produce), irrigation water must be filtered if it is to be used in a drip system. Chemical precipitates are usually associated with groundwater sources, while organic materials are most common in surface waters.

Sand, silt, plant material, algae, fish, and other materials suspended in the water must be removed before the water enters the drip line. Any particles larger than about one-tenth the size of the emitter orifice or flow passage should be removed to prevent particles from clogging emitters and to prevent material from settling out and being deposited as sediment where the water velocity is low.

Filters may be screen or cartridge filters, sand-media filters, sand separators, or disc filters. The best filter type to use depends on the type and size of particles suspended in the water.

Screen Filters

How well a screen filter performs depends on the quantity of suspended material, the mesh size, the flowrate, and the dimensions of the screen surface. Mesh sizes commonly available range from 20 to 200, with the mesh rating referring to the number of wires per inch. The larger numbers retain the smallest particles. For most drip systems, a 200 mesh screen is recommended. Screens are made from stainless steel, nylon, or polyester. The maximum flowrate recommended with a screen filter is 200 gallons per minute per square foot of screen open area. Cleaning should take place when the pressure drop across the filter has increased by 3 to 5 psi. Screen filters may be the most economical choice for use with groundwater sources, since they are effective at removing inorganic materials, but they are not effective for use with surface waters because they can quickly become clogged by large amounts of organic matter, making frequent cleaning necessary.

Sand Media Filters

Media filters *(Figure 1)*, are tanks made of carbon steel, stainless steel, or, less frequently, fiberglass, and filled with a filtering media—usually silica sand or crushed granite. The particle size of the media is selected according to the desired degree of filtration. Contaminants are filtered from the water as the water flows down through the filtering media. An underdrain of either epoxy cake or slotted or perforated screen at the bottom of the tank retains the media during filtration. Carbon steel tanks are the most common media filters, but may be subject to corrosion under certain conditions. The more expensive stainless steel tanks are lighter and have a longer service life than the carbon steel tanks.

Media filters have a greater filtering capacity than screen filters and can be used to remove both organic contaminants and inorganic contaminants, which makes them suitable for use with surface waters. Usually several media filters are placed in tandem so that as one filter is being backflushed, the other filter(s) can continue to provide clean water for the backwashing and to the irrigation system. Additional filter tanks can be added if increased filtration capacity is needed. A backup screen filter should be placed downstream to catch any sand escaping the media filter, either from routine operation or from failure of the filter's underdrain.

How well a sand media filter performs depends on the water quality, the type and size of the media, the flowrate, and on how much the pressure drops as the water flows through the sand. If the filter is too small, the pressure will drop too much, making frequent flushing necessary. *Table 1* specifies the relationship between sand media size and screen mesh designation.

Table 1. Sand media size and screen mesh designation.

Sand No.	Effective Sand Size (inches)	Screen Mesh Designation
8	0.059	70
11	0.031	140
16	0.026	170
20	0.018	230
30	0.011	400

The recommended flowrate for sand media filters is 15 to 25 gallons per minute per square foot of filter surface area. Higher flowrates can be used where the water contains less than 10 ppm of suspended material. If the water has 100 ppm or more of suspended material, lower flowrates should be used to avoid the need for very frequent backflushing. At 25 gallons per minute per square foot, the maximum size of particles retained is about one-twelfth the average size of sands 8 and 11 and about one-fifteenth the average size of sand sizes 16, 20, and 30.

Where sediment load is high it may be advantageous to pre-screen the water. If sand is present in the water, a vortex or cyclone sand separator *(Figure 2)* can be used to remove large particles that are heavier than water prior to filtration.

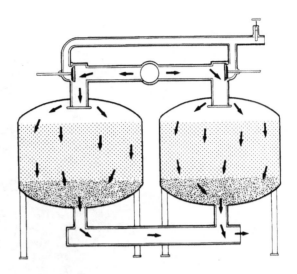

Figure 1. Media filter with flow in the filtration mode.

Sand media filters should be backflushed when the pressure drop reaches about 10 psi. Recommended flowrates for backflushing depend on the media size. Flowrates of 10-15 gpm/ ft² are suggested for # 30 and #20 media and flowrates of 20 to 25 gpm/ft² are suggested for #16 and #11 media.

Sand Separators

Centrifugal sand separators *(Figure 2)* swirl the water, creating a centrifugal force that causes sand and other heavy particles to settle out. These devices can remove up to 98% of particles larger than the equivalent of 200 mesh. Centrifugal separators are sized according to flowrate and operate at a 5 to 10 psi pressure drop. They are not effective at removing organic material.

Disc Filters

Disc filters consist of a stack of discs, each containing a series of microscopic grooves *(Figure 3)*. Water is filtered as it flows through the grooves. One disk filter manufacturer offers mesh sizes ranging from 40 to 600 mesh.

Most disk filters must be cleaned manually, but there are automatic backflush disk filters available. Disk filters separate during backflushing and require less water volume for backflushing than do sand media filters. One manufacturer specifies a flowrate of 9 to 18 gpm for each filter unit. A backflushing pressure of 50 psi is recommended. A separate booster pump may be required for backflushing.

Disk filters are sometimes promoted as an alternative to sand-media filters for filtering organic matter such as algae. Research has shown that disk filters are as effective as sand-media filters in removing organic matter. However, they tend to clog more rapidly than do sand-media filters. Thus, only disk filters with automatic backflashing capabilities should be used as an alternative to sand-media filters.

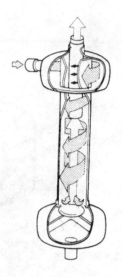

Figure 2. Centrifugal separator.

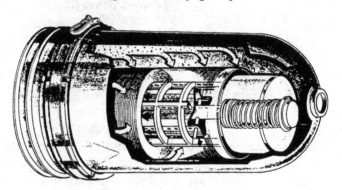

Figure 3. Disc filter.

Gravity Filters

Gravity filters can be used with water having low to medium concentrations of suspended solids (10 to 100 ppm). With these devices, the water flows through a screen by the force of gravity. The screen, with mesh size ranging from 100 to 200 mesh, is cleaned by the horizontal flow of water across the screen and by rotating cleaning jets. One manufacturer requires at least 30 psi for operating the cleaning jets. Contaminants are collected in a trash tank where they must be removed periodically by hand. Water must be re-pressurized following use of a gravity-flow screen filter so pumping/energy costs should be considered.

Suction Screen Filters

Suction screen filters are rough filters used for prescreening and are placed at the inlet of the pump intake pipe when surface water is being pumped. The filters should be used where concentrations of organic materials are moderate. Mesh sizes range from 10 to 30. These systems use rotating water jets inside the screen for self-cleaning. Pressure for the jets is supplied by the discharge side of the pump. Flowing water is needed to remove trash jetted off the screen.

Table 2 on the following page presents filtration guidelines.

Table 2. Filtration Guidelines.

Flowrate	Concentrations		Filtration*
	Organic	Inorganic	
less than 50 gpm	L	L	A
	L	M	C +A
	L	H	C + A
	M	L	D or A
	M	M	C + D, or C + A
	M	H	C + D, or C + A
	H	L	D or A
	H	M	C + D or C + A
	H	H	C + D or C + A
50 - 200 gpm	L	L	A
	L	M	C + A
	L	H	C + A
	M	L	B or E
	M	M	C + B or C + D
	M	H	C + B, C + D or C + E
	H	L	B, D or E
	H	M	C + D, C + D or C + E
	H	H	C + B, C + D or C + E
more than 200 gpm	L	L	A
	L	M	C + A
	L	H	C + F or E, C + A or C + E
	M	L	B or E
	M	M	C + B or C + E
	M	H	C + B or C + E
	H	L	B or E
	H	M	C + B or C + E
	H	H	C + B or C + E

L = less than 5 ppm
M = 5 - 50 ppm
H = more than 50 ppm

A = screen/disk filter
B = suction screen filter
C = centrifugal separator
D = gravity filter
E = sand media filter

* Letter sequence indicates the sequence of the filters: C + E means a centrifugal separator followed by a sand media filter

References

Boswell, M.J. 1990. *Micro-irrigation design manual.* Hardie Irrigation.

Center for Irrigation Technology. 1986. *Micro-irrigation methods and materials update.* California Agricultural Technology Institute.

Keller, J. and R.D. Bliesner. 1990. *Sprinkler and trickle irrigation.* Van Nostrand Beinhold, New York.

Bruce, D.A. 1985. "Filtration and analysis and applications." In: *Drip/trickle irrigation in action, Proceedings of the third international drip/trickle irrigation congress,* Fresno, CA. 18-21 November.

Injection Devices

By Larry Schwankl, UC Irrigation Specialist

Chemicals are often injected through irrigation systems, particularly micro-irrigation (drip and micro-sprinkler) systems. This process, known as *chemigation,* allows a manager to apply chemicals at any time without the need for equipment in the field. Chemigation both increases the efficiency of chemical application — resulting in decreased chemical use and cost — and reduces the hazard to those handling and applying the chemicals. It is also potentially less harmful to the environment, compared to air applications, for instance, which may allow chemical wind drift. However, chemigation can still cause environmental damage, particularly when the chemicals injected move readily with the irrigation water. Over-irrigation resulting in deep percolation can contaminate groundwater when a mobile chemical is injected.

Many different substances can be injected through irrigation systems, including chlorine, acid, fertilizers, herbicides, micronutrients, nematocides, and fungicides. Of these, chlorine, acid, and fertilizers are the substances most commonly injected. Chlorine or acid injection is used in micro-irrigation systems to prevent clogging caused by biological growths (algae and bacterial slimes) and chemical precipitation (particularly calcium carbonate).

Irrigators wishing to inject chemicals have a variety of injection equipment from which to choose, including differential pressure tanks, venturi devices, and positive displacement pumps.

Differential Pressure Tanks

Differential pressure tanks, often referred to as "batch tanks," are the simplest of the injection devices. The inlet of a batch tank is connected to the irrigation system at a point of pressure higher than that of the outlet connection. This pressure differential causes irrigation water to flow through the batch tank containing the chemical to be injected. As the irrigation water flows through the batch tank, some of the chemical goes into solution and passes out of the tank and into the downstream irrigation system. Because the batch tank is connected to the irrigation system, it must be capable of withstanding the operating pressure of the irrigation system.

While relatively inexpensive and simple to use, batch tanks have the disadvantage that as irrigation continues, the chemical mixture in the tank becomes more and more diluted, decreasing the concentration in the irrigation water *(Figure 1)*. Some injection equipment available uses a modified batch tank principle by diverting only a controlled portion of the irrigation water through the injector - allowing better control of the injection rate. If a set amount of a chemical, such as fertilizer, is to be injected and concentration during the injection is not critical, batch tanks may be appropriate to use. But if the chemical concentration must be kept relatively constant during injection, batch tanks should not be used.

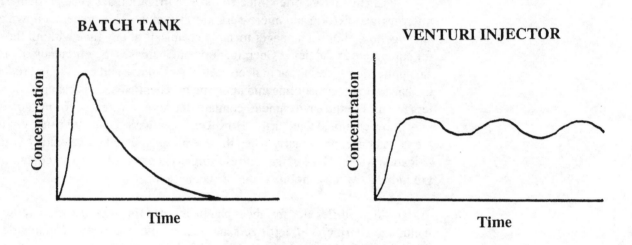

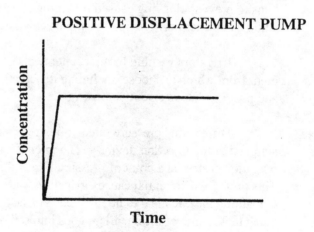

Figure 1. Chemical concentration levels during injection using a batch tank,
venturi injector, and positive displacement pump.
(*Source:* I. Bosconer. "How Irrigation Lines Can Serve Double Duty." *Agricultural Engineering* May/June 1987.)

Venturi Devices

Venturi devices (Figure 2) — often referred to as "mazzei injectors"— consist of a constriction in a pipe's flow area, resulting in a negative pressure or suction at the throat of the constriction. "Mazzei" is simply a trade name for a particular brand of venturi injector. Venturi injectors are also available from other manufacturers.

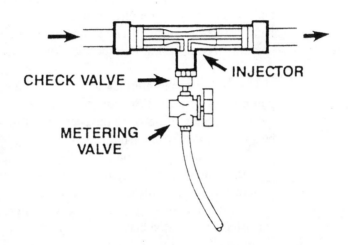

CHECK VALVE →   ← INJECTOR

METERING VALVE →

Figure 2. Venturi device.

The venturi injector is frequently installed across a valve or other point where between 10 and 30 percent of the system pressure is lost. Because of friction in the venturi, this pressure differential is necessary to allow flow through the venturi injector. This means that the inlet of the injector must be at a pressure 10 to 30 percent higher than the outlet port. Because of these significant pressure losses, the injector should be installed parallel to the pipeline so that flow through the injector can be turned off with a valve when injection is not occurring. The injection rate of a venturi device is determined by the size of the venturi and the pressure differential between inlet and outlet ports. Injection rates as high as 700 gallons per hour are possible with large venturi devices.

Venturi devices are inexpensive and relatively simple to operate, but do not inject chemicals at as constant a rate as positive displacement pumps*(Figure 1)*. Injecting with venturi devices may be sufficiently accurate for applications such as fertilizer injection, however.

Positive Displacement Pumps

Positive displacement pumps are piston or diaphragm pumps that inject at precise rates. The pumps are powered by either electricity or gasoline or are driven by water. The water-driven pumps can be installed in locations that lack power. When a constant and precise injection concentration is needed, positive displacement pumps are preferable *(Figure 1)*.

Positive displacement pumps are the most expensive of the injection devices, with costs for electric pumps running $750 or more.

Safety and Environmental Protection

Contamination can occur if: (1) the irrigation water pumping plant shuts down while the injection unit continues to operate; or (2) irrigation water is allowed to flow back through the injection unit and into the chemical storage tank; causing the tank to overflow.

Backflow prevention devices and interlocks between the injection devices and pump(s) are necessary, and often legally required, to prevent contamination. Legally required equipment should be determined by contacting your local regulatory agency. The following are safety and contamination prevention devices frequently required for injection systems.

• Electrical interlock between the pump and the injector. If the irrigation pumping plant stops, this shuts down the injector.
• Shut off between the chemical storage tank and the injector. This may be a normally-closed solenoid valve.
• A check valve on the injection line just prior to it entering the irrigation system. This prevents irrigation water from flowing back into the injection system and overflowing the storage tank.
• Irrigation system check valve (single or double available-check local requirements) located between the injection point and the pump. This prevents chemicals from flowing back through the irrigation system and contaminating the water source.
• A vacuum relief valve is often installed between the check valve and the pump. This prevents a vacuum from forming, which could cause backwards flow of chemicals to the water source, when the pump is shut down and water leaves the irrigation system.
• An automatic cutoff sensor may also be installed in the irrigation system. This sensor will automatically shut down power to the injector if system pressure drops too low - an indication that the pump has shut down.

All chemicals should be stored in separate tanks, usually made of polyethylene or fiberglass. Mixing any chemicals together should be avoided. Some mixing of chemicals, such as mixing acid and chlorine together, can produce highly toxic by products; in this case chlorine gas.

Operating the System

Converting to Micro-irrigation
By Larry Schwankl, UC Irrigation Specialist

Converting an established orchard or vineyard from surface flood or sprinkler irrigation to micro-irrigation offers the grower a number of advantages—including making irrigation more uniform, improving efficiency by maximizing the use of water, reducing labor needs by automating the system, reducing the cost of controlling summer weeds, and using the system to apply fertilizer. But it also requires attention to important additional issues beyond the basics of design, operation, and maintenance discussed elsewhere in this handbook.

Restricted Root Zone

The first of those issues is the difference in root patterns that develop under surface flood or sprinkler systems and those that develop under a micro-irrigation system. Under a surface flood or sprinkler system, the root zone mirrors the full coverage of the irrigation system, while under a micro-irrigation system, the root zone reflects the more restricted wetted area.

Switching to a micro-irrigation system therefore requires the tree or vine to adapt its root system to enable it to draw water from the smaller zone wetted by the dripper or micro-sprinkler. For this reason, converting to a micro-sprinkler system may be less problematic than converting to other micro-irrigation systems, since micro-sprinklers generally wet a larger soil area than drip systems.

Because of the need for the plant roots to adapt to the new system, it is particularly important that the micro-irrigation system deliver sufficient water—enough to provide a wetted area not less than 40% of the orchard floor—and that the system be able to deliver the water within a 16 to 18-hour period each day during peak water demand periods. Some growers prefer that the system provide a wetted area of up to 60% of the orchard floor.

Frequent Irrigation

Equally important in making the change to micro-irrigation is realizing the need for the crop to be irrigated more often. During peak water use periods, tree crops must often be irrigated daily with a drip irrigation system, while flood or sprinkler irrigation systems might allow an interval of a week or more between irrigations. The water manager should know the application rate of the micro-irrigation system in inches or gallons per hour in order to be certain that the system operates long enough and often enough to meet the needs of the crop. *(See "How Much Water is Being Applied," page 71 of this handbook.)*

Transition Period

Following are measures that can ease the transition of the orchard or vineyard to the new micro-irrigation system:

• go into the growing season with a soil profile full of water. If this soil moisture is not provided by winter rainfall, begin irrigations early to provide a "water bank account" for the crop to draw on if it is needed.

• slightly over-irrigate the crop during the first growing season (but avoid extensive over-irrigation)

• if it is still in place, apply one or two irrigations with the old system during the first season after converting to micro-irrigation.

Cover Crops

Most drip or micro-sprinkler systems are not well-suited to the needs of cover crops. If a cover crop is required, the grower should weigh that requirement in considering whether to convert to micro-irrigation.

Wetting Patterns
By Blaine Hanson, UC Irrigation and Drainage Specialist

Compared to other irrigation methods, micro-irrigation wets a relatively small portion of the soil surface because it applies water from a point source. Micro-sprinklers wet a larger area — although one still smaller than that wetted by sprinkler or border irrigation.

Under micro-irrigation, the soil is wettest near the emitter or micro-sprinkler and progressively drier farther away from the emitter. This feature has the advantage of keeping surface evaporation and weed growth down, but the disadvantage of limiting the wetting of the potential root zone.

Figure 1 on the following page illustrates the wetting patterns that would occur during infiltration using a surface drip system in two different soil types — loam and sand — and at two different discharge rates — 1 gallon per hour and 4 gallons per hour. The same volume of water is applied in each case. The solid line at the soil surface represents the area of water ponded on the surface.

At the 1 gph discharge rate, the lateral water movement after three hours of infiltration was much greater — and the area of ponded water larger — in the loam than in the sand, but the water had infiltrated the sand much more deeply than the loam.

At the 4 gpm discharge rate, the ponded area at the surface was much larger than at the lower discharge rate, resulting in much more lateral movement of the water and shallower infiltration.

In sandy soil, then, lateral wetting is much less than in fine-textured soil, because the lateral flow of water in sand is relatively slight. But these tests show that lateral wetting can be increased by increasing the emitter discharge rate, thereby causing the area of ponded water to also increase. The area of ponded water will increase until the water infiltration equals the emitter discharge rate.

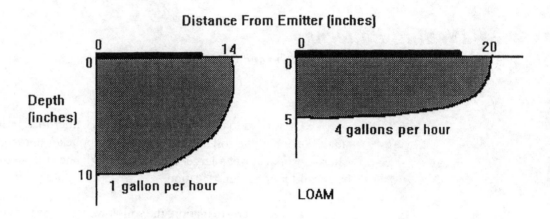

Figure 1a. *Wetting pattern during infiltration using a surface drip system in loam soil at discharge rates of 1 gallon per hour and 4 gallons per hour.*

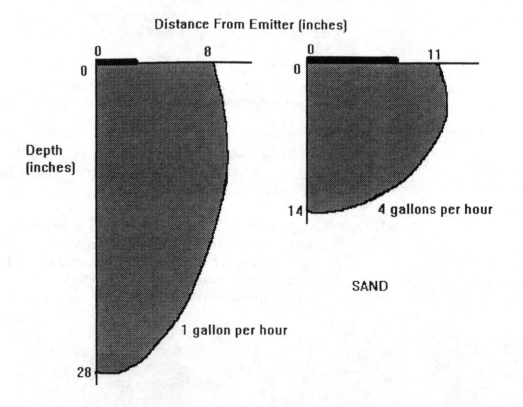

Figure 1b. *Wetting pattern during infiltration using a surface drip system in sandy soil at discharge rates of 1 gallon per hour and 4 gallons per hour.*

Figure 2 illustrates water distribution patterns under surface and subsurface drip irrigation with water applied at a rate equal to 100 percent of evapotranspiration. Reflecting a continuing pattern of infiltration and plant extraction, the figure shows that soil moisture content is highest near the emitter and remains relatively constant at depths directly beneath the emitter, but gradually decreases with horizontal distance from the emitter.

With subsurface drip tape buried about 18 inches deep, *Figure 2* shows that soil moisture is highest near the tape and relatively constant below the tape, but decreases both above the buried tape and with lateral distance from it.

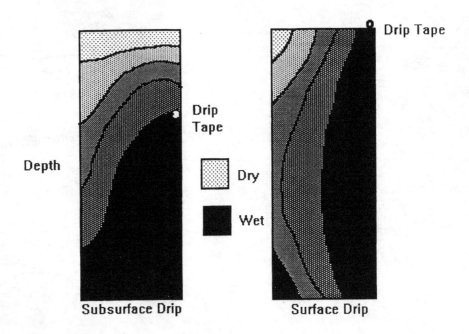

Figure 2. Water distribution patterns under surface and subsurface drip irrigation with water appied at a rate equal to 100 percent of evapotranspiration.

Figures 3 and 4 compare wetting patterns of drip emitters and micro-sprinklers. They show that drip emitters may wet only a small volume of soil, while micro-sprinklers wet a much larger area. Under micro-sprinklers, wetted diameters often range from 10 to 20 feet.

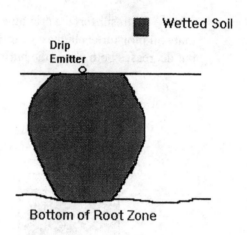

Figure 3. Wetting pattern of drip emitter.

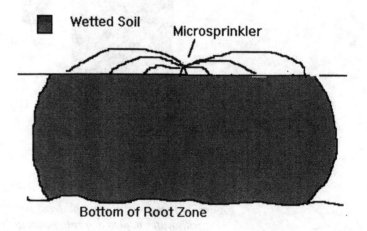

Figure 4. Wetting pattern of micro-sprinkler.

How Often to Irrigate

By Larry Schwankl, UC Irrigation Specialist, and
Terry Prichard, UC Water Management Specialist

Determining how often to irrigate is central to managing a micro-irrigation system. The following factors should be taken into account in making this decision:

Factors to Consider

• *Design capacity of the system.* Decisions about irrigation frequency and duration for a micro-irrigation system are made at the time the system is designed. Usually the design is based on peak crop water use requirements—that is, on the amount of water required during the time a mature crop needs the most water. To minimize hardware costs, most micro-irrigation systems are designed to operate for long periods per day.

To determine how many hours per day the irrigation system should operate, find the irrigation system application rate (inches/hour) *(See "How Much Water is Being Applied?" page 71)* and compare it to the crop water demands (inches/day). Remember to factor in extra operating time to compensate for irrigation inefficiencies. For an irrigation system operating at 80% efficiency, allow 25% more operating time. When crop water use is not at its peak, the system will not have to operate as long or as often.

• *Soil Moisture*. A micro-irrigation system makes it possible to irrigate frequently and to keep the soil moisture level high, which provides optimum growing conditions for the plant with little stress to threaten yield or quality.

Most growers find that sandier soils hold less water than heavier soils and therefore must be irrigated more often. Remember that the main purpose of each irrigation is to replace the soil water used since the last irrigation, so that less-frequent irrigations must be longer in duration.

• *Wetted area.* Micro-irrigation wets only a portion of the crop root zone. While supplying adequate water to even a small portion of the root zone may be adequate for successful plant growth, other considerations, such as tree stability, nutritional requirements, and the need for adequate reserve moisture to mitigate high environmental demand suggest that wetting a larger area (40 to 50% of the soil area) may be advantageous. This is accomplished more easily with micro-sprinklers than with drip emitters. Less frequent irrigations of longer duration tend to promote more lateral spreading of water.

• *Off-peak pumping.* Micro-irrigation systems also make it possible — and advantageous—to pump during off-peak energy periods, thereby reducing energy costs. Since this practice requires that a specified time window be avoided, the remaining hours available for operation influence irrigation frequency.

Recommendations

Both scientific research and actual grower experience has shown the following irrigation frequencies to be optimum during periods of high crop water use:

• *Drip emitter systems:* one to three days between irrigations.

• *Micro-sprinkler systems:* two to four days between irrigations.

How Much Water is Being Applied?
By Larry Schwankl, UC Irrigation Specialist

Determining the application rate of micro-irrigation systems can be confusing because irrigation scheduling and crop water use information is usually presented in inches per day (in/day), while discharge from micro-irrigation emitters is measured in gallons per hour (gph). The following may be helpful in determining required operating times for micro-irrigation systems.

The water use of the crop and the application rate of the emission device(s) determines how long drip and micro-sprinklers should be operated.

Step 1

Step 1 in determining the required operating time is to convert the crop water use information (usually available in inches per day), to gallons per day of plant water use. The following formula may be used (or see *Table 1*):

$$\begin{array}{lcccccc} \text{Water use} & & \text{Plant} & & \text{Crop water} & & \\ \text{by the plant} & = & \text{spacing} & \text{X} & \text{use} & \text{X} & 0.623 \qquad (1) \\ \text{(gal/day)} & & \text{(ft}^2\text{)} & & \text{(in/day)} & & \end{array}$$

Tree crop spacing = 20 ft. x 20 ft. = 400 ft^2

Crop Water Use = 0.3 in/day

Example

Water Use by
   the Plant = 400 ft^2 x 0.3 in/day x 0.623

          = 75 gal/day

or refer to *Table 1*.

Table 1. Crop water use (gal/day) for various plant spacings and crop water use (in/day).

Crop water use (in/day)

Crop Spacing (ft^2)	0.05	0.1	0.15	0.2	0.25	0.3	0.35	0.4
100	3	6	9	12	16	19	22	25
200	6	12	19	25	31	37	44	50
400	12	25	37	50	62	75	87	100
600	19	37	56	75	93	112	131	150
800	25	50	75	100	125	150	174	199
1000	31	62	93	125	156	187	218	249
1200	37	75	112	150	187	224	262	299
1400	44	87	131	174	218	262	305	349
1600	50	100	150	199	249	299	349	399
1800	56	112	168	224	280	336	392	449
2000	62	125	187	249	311	374	437	498
2200	69	137	206	274	343	411	480	548
2400	75	150	224	299	374	449	523	598

Crop spacing (ft^2) = row spacing (ft) x plant spacing (ft)

Step 2

Step 2 is to determine the application rate of the irrigation system in gallons per hour (gal/hr). For both drip emitters and micro-sprinklers, this requires determining: (1) the number of emission devices per plant, and (2) the discharge rate per emission device (gal/hr/emitter):

| Application Rate (gal/hr) | = | Number of Emission Devices | X | Discharge Rate per Emission Device (gal/hr/emitter) | (2) |

Example

Drip emitters: 4 drip emitters per tree

Discharge rate per emitter = 1 gal/hr

Application rate = 4 emitters/tree x 1 gal/hr per emitter (gal/hr)
            = 4 gal/hr.

Micro-sprinklers: 1 micro-sprinkler per tree

Discharge rate per micro-sprinkler = 12 gal/hr.

Application rate (gal/hr) = 1 micro-sprinkler/tree x 12 gal/hr per
                micro-sprinkler
                = 12 gal/hr.

Step 3

Step 3 is to determine the irrigation system operation time in hours per day. This requires the crop water use (determined in **Step 1**), and the application rate (determined in **Step 2**). The following formula may be used (or see *Table 2*):

$$\text{Hours of operation per day} = \frac{\text{Crop water use (gal/day)}}{\text{Application rate (gal/hr)}} \qquad (3)$$

Table 2. Hours of operation per day for various application rates (hrs/day) and crop water use (gal/day).

Application rate (gal/hr)

Crop Water Use (gal/day)	1	2	4	6	8	10	12	14	16	18	20
5	5.0	2.5	1.3								
10	10.0	5.0	2.5	1.7	1.3	1.0					
15	15.0	7.5	3.8	2.5	1.9	1.5	1.3	1.1			
25		12.5	6.3	4.2	3.1	2.5	2.1	1.8	1.6	1.4	1.3
50			12.5	8.3	6.3	5.0	4.2	3.6	3.1	2.8	2.5
75			18.8	12.5	9.4	7.5	6.3	5.4	4.7	4.2	3.8
100				16.7	12.5	10.0	8.3	7.1	6.3	5.6	5.0
125				20.8	15.6	12.5	10.4	8.9	7.8	6.9	6.3
150					18.8	15.0	12.5	10.7	9.4	8.3	7.5
175					21.9	17.5	14.6	12.5	10.9	9.7	8.8
200						20.0	16.7	14.3	12.5	11.1	10.0
225						22.5	18.8	16.1	14.1	12.5	11.3
250							20.8	17.9	15.6	13.9	12.5
275							22.9	19.6	17.2	15.3	13.8
300								21.4	18.8	16.7	15.0
325								23.2	20.3	18.1	16.3
350									21.9	19.4	17.5
375									23.4	20.8	18.8
400										22.2	20.0
425											21.3
450											22.5
475											23.8

Example

Drip emitters:
   Crop water use (gal/day) = 75 gal/day (Step 1)
   Application rate (gal/hr) = 4 gal/hr (Step 2)
   Hours of operation per day = 75 gal/day ÷ 4 gal/hr = 18.8 hrs/day

Micro-sprinklers:
   Crop water use (gal/day) = 75 gal/day
   Application rate (gal/hr) = 12 gal/hr

   Hours of operation per day = 75 gal/day ÷ 12 gal/hr = 6.3 hrs/day

Table 2 gives the same hours of operation for these examples.

Drip Tapes and Tubing

Drip tapes and tubings, placed on the soil surface or subsurface, are most often used for irrigating row crops. Determining daily operation times for these systems is somewhat more complicated than for drip emitters and micro-sprinklers, but follows a similar three-step process. The discharge rate of drip tapes and tubings is usually given in gallons per minute per 100 feet of material (gal/min per 100 ft).

Step 1

Step 1 is to determine the crop water use in inches per day (in/day), which is the standard measure used in evapotranspiration (ET)-based methods of irrigation scheduling.

Step 2

Step 2 is to determine the application rate of the drip tape or tubing in inches per hour (in/hr). *Table 3* can be used to make this determination if the row spacing and the irrigation system application rate (gal/min per 100 feet) are known.

If the drip tape lateral spacing you need is not in Table 3, the following formula can be used to calculate the application rate.

$$\text{Application Rate (in/hr)} = \text{Drip tape discharge in (gpm/100 ft)} \div \text{Spacing between tape laterals (ft)} \times 0.963$$

Example

Row spacing = 60 inches
Drip tape application rate = 0.5 gal/min per 100 ft.

From *Table 3:* the application rate = 0.1 in/hr.

Step 3

Step 3 is to determine the irrigation system operation time (in hours per day) necessary to satisfy the crop water needs. This requires the crop water use (determined in *Step 1*), and the application rate (determined in *Step 2*). The following formula may be used (or see *Table 4*):

$$\text{Hours of operation per day} = \frac{\text{Plant water use (in/day)}}{\text{Application rate (in/hr)}} \qquad (4)$$

Table 3. Application rate (in/hr) of drip tapes and tubings
for various flowrates and spacings.

					Flowrate (gal/min per 100 ft)					
		0.1	0.15	0.2	0.25	0.3	0.35	0.4	0.45	0.5
	12	0.10	0.14	0.19	0.24	0.29	0.34	0.39	0.43	0.48
	14	0.08	0.12	0.17	0.21	0.25	0.29	0.33	0.37	0.41
	16	0.07	0.11	0.14	0.18	0.22	0.25	0.29	0.32	0.36
	18	0.06	0.10	0.13	0.16	0.19	0.22	0.26	0.29	0.32
	20	0.06	0.09	0.12	0.14	0.17	0.20	0.23	0.26	0.29
	22	0.05	0.08	0.11	0.13	0.16	0.18	0.21	0.24	0.26
	24	0.05	0.07	0.10	0.12	0.14	0.17	0.19	0.22	0.24
	26	0.04	0.07	0.09	0.11	0.13	0.16	0.18	0.20	0.22
	28	0.04	0.06	0.08	0.10	0.12	0.14	0.17	0.19	0.21
	30	0.04	0.06	0.08	0.10	0.12	0.13	0.15	0.17	0.19
	32	0.04	0.05	0.07	0.09	0.11	0.13	0.14	0.16	0.18
Row	34	0.03	0.05	0.07	0.08	0.10	0.12	0.14	0.15	0.17
spacing	36	0.03	0.05	0.06	0.08	0.10	0.11	0.13	0.14	0.16
(in)	38	0.03	0.05	0.06	0.08	0.09	0.11	0.12	0.14	0.15
	40	0.03	0.04	0.06	0.07	0.09	0.10	0.12	0.013	0.14
	42	0.03	0.04	0.06	0.07	0.08	0.10	0.11	0.12	0.14
	44	0.03	0.04	0.05	0.07	0.08	0.09	0.11	0.12	0.13
	46	0.03	0.04	0.05	0.06	0.08	0.09	0.10	0.11	0.13
	48	0.02	0.04	0.05	0.06	0.07	0.08	0.10	0.11	0.12
	50	0.02	0.03	0.05	0.06	0.07	0.08	0.09	0.10	0.12
	52	0.02	0.03	0.04	0.06	0.07	0.08	0.09	0.10	0.12
	54	0.02	0.03	0.04	0.05	0.06	0.07	0.09	0.10	0.11
	56	0.02	0.03	0.04	0.05	0.06	0.07	0.08	0.09	0.10
	58	0.02	0.03	0.04	0.05	0.06	0.07	0.08	0.09	0.10
	60	0.02	0.03	0.04	0.05	0.06	0.07	0.08	0.09	0.10

Example

Crop water use = 0.3 in/day
System application rate = 0.1 in/hour
Hours of operation per day = 0.3 in/day ÷ 0.1 in/hour = 3 hrs/day

Table 4 gives the same operation time for this example.

Table 4. Operation time (hrs/day) for various application rates (in/hr) and crop water use (in/day).

Plant water use (in/day)	Application rate (in/hr): 0.05	0.1	0.15	0.2	0.25	0.3	0.35	0.4	0.45	0.5
.05	1.0	0.5	0.3	0.3	0.2	0.2	0.1	0.1	0.1	0.1
0.1	2.0	1.0	0.7	0.5	0.4	0.3	0.3	0.3	0.2	0.2
0.15	3.0	1.5	1.0	0.8	0.6	0.5	0.4	0.4	0.3	0.3
0.2	4.0	2.0	1.3	1.0	0.8	0.7	0.6	0.5	0.4	0.4
0.25	5.0	2.5	1.7	1.3	1.0	0.8	0.7	0.6	0.6	0.5
0.3	6.0	3.0	2.0	1.5	1.2	1.0	0.9	0.8	0.7	0.6
0.35	7.0	3.5	2.3	1.8	1.4	1.2	1.0	0.9	0.8	0.7
0.4	8.0	4.0	2.7	2.0	1.6	1.3	1.1	1.0	0.9	0.8

Calculating Pressure Loss in Mainlines and Submains
By Larry Schwankl, UC Irrigation Specialist

Pressure, which is defined as force over an area, is lost from a micro-irrigation system as a result of water moving uphill or from friction as water moves through pipes, valves, filters, flowmeters, and other elements in the system.

Pressure is usually expressed either in pounds per square inch (psi) or in feet of head. The relationship between psi and feet of head is as follows:

$$1 \text{ psi } = 2.31 \text{ feet of head} \qquad (1)$$

Micro-irrigation emission devices (such as drippers and micro-sprinklers) are designed to operate within a certain pressure range. If an operating pressure at the emitters has been chosen, pressure variation from elevation differences is known, and all pressure losses from friction are quantified, it is possible to determine the pressure required at the head of the system. This pressure is usually supplied by a pump.

The pipelines (mainlines and submains) in micro-irrigation systems are usually made of PVC pipe, with Class 160 or Schedule 40 pipe being common. Pressure loss from friction in these pipelines depends on the pipe's length, its inside diameter, its roughness or resistance to flow, and the flowrate. Pressure loss from friction in PVC pipe (psi/100 feet of pipe) can be determined from *Tables 1- 4* in Appendix I. *Table 1*, below, is a key to the tables.

Table 1. Key to Pressure Loss Tables 1-4, Appendix I.

Table Number	Type of PVC Pipe	Size	Flow Rates (gpm)
1	Class 160	1/2" - 2"	0.1 - 44
2	Class 160	2" - 8"	10 - 1725
3	Schedule 40	1/2" - 2"	0.1 - 44
4	Schedule 40	2" - 8"	10 - 1800

Example

The pressure loss from friction is given in psi/100 feet of pipeline. As an example, a flowrate of 40 gpm in two-inch, Class 160 PVC pipe would have a pressure loss from friction of 0.98 psi/100 feet of pipe *(Table 1, Appendix I).*

If the system has multiple outlets at regular spacings along mains and submains, the flowrate downstream from each of the outlets will be effectively reduced. Since the flowrate affects the amount of pressure loss, the pressure loss in such a system would be only a fraction of the loss that would occur in a pipe without outlets. *Table 1,* below gives the fractional values (frequently called the "F" value) of a system with multiple outlet pipes. A pipe with ten regularly spaced outlets, for example, would have 0.385 times the pressure loss of an equivalent pipe with no outlets.

Pressure Loss in Fittings

Because of friction, pressure is lost whenever water passes through fittings, such as tees, elbows, constrictions, or valves. The magnitude of the loss depends both on the type of fitting and on the water velocity (determined by the flow rate and fitting size). Pressure losses in major fittings such as large valves, filters, and flow meters, are readily determined from information supplied by the manufacturer. If a detailed accounting is required of minor pressure losses in fittings such as tees and elbows, the reader should refer to a hydraulics hand-book. Minor losses can also be aggregated into a friction loss safety factor (10 percent is frequently used) over and above the friction losses in pipelines, filters, valves, and other elements, which are more easily accounted for.

Table 2. Fractional ("F") values for reducing pressure loss in multiple outlet pipes as compared to pipes with no outlets.

Number of Outlets	"F"	Number of Outlets	"F"
1	1.000	16	.365
2	.625	17	.363
3	.518	18	.361
4	.469	19	.360
5	.440	20	.359
6	.421	21	.357
7	.408	24	.355
8	.398	26	.353
9	.391	28	.351
10	.385	30	.350
11	.380	35	.347
12	.376	40	.345
13	.373	50	.343
14	.370	100	.338
15	.367	inf.	.333

Calculating Pressure Loss in Lateral Lines
By Larry Schwankl, UC Irrigation Specialist

Pressure losses in the micro-irrigation system laterals — that is, the pipes in which the emitters are installed (usually polyethylene "poly" tubing) — can be caused by water moving uphill or by friction in the lateral lines.

Pressure, which is defined as force over an area, is usually expressed either in pounds per square inch (psi) or in feet of head. Psi is converted to feet of head as follows:

$$1 \text{ psi } = 2.31 \text{ feet of head} \qquad (1)$$

Measuring the friction-caused pressure losses in lateral lines is important in order to: (1) determine how long the laterals should be for optimum application uniformity and (2) calculate the lateral line component of the total pressure loss from friction in all parts of the system, including mainlines/submains, valves, filters, flow meters, and by other factors such as changes in elevation. Once all pressure losses have been measured and the desired emitter operating pressure chosen, the pressure requirements at the pump can be determined. If there are still significant pressure differences within the system despite careful system design, pressure-compensating emitters may have to be used to achieve optimum application uniformity.

Friction-caused pressure losses in the lateral lines are influenced by the diameter and length of the poly tubing and by the lateral line flowrate. The lateral line flowrate can be determined from the spacing and discharge rate of the emitters installed in the lateral.

Where lateral lines are of a single size, an easy way to calculate pressure loss from friction is to determine the specific discharge rate (SDR) of the lateral line and then refer to friction loss charts. The SDR (in gph/ft) can be determined by dividing the emitter discharge rate by the emitter spacing. For example, a lateral line with emitters of 1 gph, spaced four feet apart would have a SDR of 1 gph ÷ 4 feet = 0.25 gph/ft.

Figures 1 - 7 in *Appendix II* can be used to determine lateral line pressure losses for polyethylene hose of varying sizes. *Table 1* is a key to the figures.

Table 1. Key to Figures 1-7 (Appendix II), showing discharge rates for polyethylene hose.

Figure No.	Polyethylene tubing size (mm)
1	10
2	13
3	15
4	16 (inside diameter)
5	16 (outside diameter)
6	20
7	26

Example

Once the SDR for a particular lateral line size is known, the pressure loss (feet of head) can be determined for any lateral length. For example, for a 16-mm poly hose (inside diameter), with a SDR of 0.25 gph/ft. (a 1-gph emitter every four feet), pressure losses for various lateral lengths can be determined from *Figure 5 (Appendix II)*. *Table 2* shows examples from *Figure 5 (Appendix II)*.

Table 2. Examples of pressure loss from inlet to tail end for various lateral lengths, from Figure 5 (Appendix II).

Length of Poly hose (ft)	Pressure loss	
	(feet of head)	*(psi*)*
200	0.5	0.22
300	2.5	1.1
400	5.5	2.4
500	10.0	4.3
600	16.5	7.1

* psi = feet of head ÷ 2.31

Source: Michael J. Boswell, 1986. *Micro-Irrigation Design Manual.* Hardie Irrigation.

Fertigation

By Larry Schwankl, UC Irrigation Specialist, and
Terry Prichard, UC Water Management Specialist

Fertigation is the common term for injecting fertilizers or soil amendments through the irrigation system. Micro-irrigation systems are well suited to fertigation because of the frequency of operation and because water application can be easily controlled by the manager. Applying fertilizers through a micro-irrigation system:

- makes fertilizer distribution relatively uniform
- allows flexibility in timing fertilizer application
- reduces the labor required for applying fertilizer, compared to other methods
- allows less fertilizer to be used compared to other fertilization methods
- can lower costs

Safety and Environmental Protection

Contamination can occur if (1) the irrigation water pumping plant shuts down while the injection equipment continues to operate, causing contamination of the water source or unnecessary amounts of fertilizer to be injected into the irrigation system, or (2) the injection equipment stops while the irrigation system continues to operate, causing the irrigation water to flow into the chemical supply tank and overflow onto the ground.

Backflow prevention devices, including vacuum breakers (atmospheric and pressure types) and check values (single and double) are available. Local regulations should be followed in selecting and using these devices.

If the injection pump is electrically driven, an interlock should be installed so that the injection pump will stop if the irrigation system pump shuts down. To keep water from flowing backward into the chemical tank, a check valve or normally closed solenoid valve, can be installed in the injection line. If an electrical solenoid valve is used, it should be connected to the injector pump and interlocked with the irrigation pump.

Fertilizer Solubility

In order to be injected, fertilizers must be soluble. Fertilizers delivered as a solution can be injected directly into the irrigation system, while those in a dry granular or crystalline form must be mixed with water to form a solution. Fertilizer materials differ widely in water solubility, with solubility depending on the physical properties of the fertilizer as well as on irrigation water quality, temperature, and pH. Dry fertilizers are mixed into a tank containing water until the granules or crystals are dissolved and the desired concentration is reached. The solution is then injected into the irrigation system.

Contaminants

Agricultural grade fertilizers and amendments are often coated to inhibit moisture absorption and to assist in material flow through machinery. These coatings and other foreign materials can cause clogging problems in the mixing tank and micro-irrigation system. The foreign material exists in the tank as sediment, falling to the bottom or as a low density scum on the surface.

To prevent problems, stock tanks should be agitated until the material is dissolved, then allowed to stand until separation occurs. Extracting solution above the sediment or skimming off the surface scum before injecting has been successful. Always filter the solution between the stock tank and injection point.

Injection Point

The injection point should be located so that the injected fertilizer and the irrigation water can become thoroughly mixed well upstream of any branching of the flow. Injecting at a point in front of the filter will reduce the chance of contaminants entering the system. The exception is injection of acids which can damage pumps and filters made of corrosion sensitive materials. A disadvantage of injecting ahead of the filters is that if the filters are backflushed during injection, injected chemical is lost with the backwash water and may be a contamination hazard. The system should be allowed to fill and come up to full pressure before injection begins. Following injection, the system should be operated to flush the fertilizer from the lines. Leaving residual fertilizer in the line may encourage clogging from chemical precipitates or organic sources such as bacterial slimes.

Injection Time

Injection of a chemical should not be done in a short, high-concentration injection. The flow velocities in drip tubing or drip tape is relatively slow and it takes water (or chemical) a significant amount of time to reach the end of the lateral. Measurements in drip tubing show that it takes 15-20 minutes for water to travel the length of a 500 foot lateral (1 gph emitters spaced 4 feet apart). In drip tape (0.22 gpm/100 feet discharge rate), it takes 30-40 minutes for water to travel to the end of a 800' lateral. To get an even distribution of the injected chemical, injection times should be significantly longer than these travel times. Constant injection rates also assist in achieving uniform chemical distribution.

Nitrogen

The fertilizer most commonly injected is nitrogen, with many soluble nitrogen sources working well in fertigation.

Following is a list of common nitrogen sources with comments about their use in fertigation:

Anhydrous ammonia or aqua ammonia. These nitrogen sources cause an increase in water pH, which may result in a precipitate if calcium or magnesium are present in the irrigation water along with comparable levels of bicarbonate. Volatilization of nitrogen (loss to the atmosphere) may also occur when anhydrous or aqua ammonia are used.

Urea. Urea is relatively soluble in the irrigation water and is not strongly held by soil particles, so it moves deeper into the soil than the ammonia products. Urea is transformed by hydrolysis into ammonium, which is then fixed to the soil particles.

Ammonium sulfate, Ammonium nitrate, and Potassium nitrate. All are relatively soluble in water and cause only a slight shift in the soil or water pH.

Calcium nitrate. Calcium nitrate is relatively soluble in water and causes only a slight shift in the soil or water pH. If the water is high in bicarbonate, however, the calcium content may lead to precipitation of calcium carbonate (lime).

Ammonium phosphate. Ammonium phosphate can also cause soil acidification. If calcium or magnesium levels are high enough in the irrigation water, precipitates may also form, which can clog the drip emitters. (See the discussion under *Phosphorus*, below, for precautions in using ammonium phosphate.)

Organic Nitrogen Sources

Organic nitrogen sources are not soluble but can be injected as a suspension through some (large orifice) micro-irrigation systems. Lines should be flushed after application to prevent biological growths feeding on the residual organic matter.

Phosphorus

Dry phosphate fertilizers are not sufficiently water soluble for injection and can cause chemical or physical precipitate clogging. Solution fertilizers containing phosphorous and various types of phosphoric acid can be injected; however, the calcium and magnesium content and the pH of the irrigation water should be considered.

If the pH of the irrigation water is not reduced to near 3.0, insoluble calcium and magnesium phosphates can be formed, clogging filters and emitters. Once insoluble dicalcium or tricalcium phosphate is formed, it cannot be resolubilzed. Take care with waters greater than 2 meq/l Ca^{+2} plus Mg^{+2}. Low rates of phosphoric acid may not sufficiently reduce the water pH to prevent precipitation. In those cases, sulfuric acid products are added to reduce the pH to overcome the buffering capacity of the water and minimize precipitation problems.

Phosphorous is quickly fixed to soil particles and does not move readily into the soil profile, but it has been found to move more easily under micro-irrigation than under conventional irrigation methods.

Potassium

Injecting potassium fertilizers usually causes few problems, but caution should be observed if potassium fertilizers are mixed with other fertilizers. Potassium, like phosphorous, is fixed by soil particles and does not move readily through the soil profile.

Potassium is usually applied in the form of potassium chloride, but for crops sensitive to chloride, potassium sulfate or potassium nitrate may be more appropriate. Potassium sulfate is not very soluble and may not dissolve well in the irrigation water.

Calculation of Injection Rates

Calculation of fertilizer injection amounts and rates requires that the fertilizer density (lbs/gallon) and the percentage of nutrient (e.g., nitrogen, potassium or phosphate) in the fertilizer be known. Table 1 lists this information for a number of nitrogen fertilizers. Most frequently, fertilizer requirements are given in terms of lbs/ac (e.g., 80 lbs of nitrogen/ac). To determine the amount of fertilizer (gallons) to be injected, the following formula can be used:

$$\frac{\text{Injection Amount}}{\text{(gallons)}} = \frac{\text{(desired lbs N/AC) x (acres) x 100}}{\text{(fertilizer density (lbs/gal)) x (\% N in fertilizer)}}$$

Since fertilizer is often purchased by weight instead of volume, the fertilizer density (shown in Table 1) can be used to convert gallons to lbs and vice versa.

To determine the injection rate (gallons per hour), the following formula can be used:

$$\frac{\text{Injection Rate}}{\text{(gal/hr)}} = \frac{\text{(desired 435 N/AC) x (acres) x 100}}{\text{[fertilizer density (lbs/gal)] x [\% N in fertilizer] x [injection time (hrs)]}}$$

Example

A walnut grower wants to apply 50 lbs/acre of nitrogen by injecting UN 32 through his drip irrigation system. The drip irrigation system services 120 trees, planted on a 28' x 28' spacing.

The acres being serviced by the drip irrigation system:

120 trees x (28' x 28'/tree) = 94,080 ft^2 = 2.16 acres
Note: 1 acre = 43,560 ft^2,   UN32 density = 11.1 lbs/gal

Using the first formula shown above:

$$\frac{\text{Injection Amount}}{} = \frac{\text{(50 lbs N/AC) x (2.16 AC) x 100}}{\text{(11.1 lbs/gal) x (32)}}$$
= 30.4 gallons of UN 32

Lbs of UN 32 = 30.4 gallons x (11.1 lbs/gal) = 337 lbs UN 32

If this amount of UN32 were to be injected during a 6-hour period, the injection rate (gph) would be:

$$\frac{30.4 \text{ gallons}}{6 \text{ hrs}} = 5.1 \text{ gph}$$

Table 1. *Liquid weights, nitrogen percentages, and lbs of nitrogen per gallon of fertilizer for various fertilizers.*

Fertilizer	Density	% Nitrogen	Lbs Nitrogen Per Gallon
ammonium nitrate	10.7	21	2.25
ammonium thiosulfate	11.0	12	1.32
aqua ammonia	7.5	20	1.50
urea ammonium nitrate (UN 32)	11.1	32	3.55
ammonium nitrate ammonia	11.7	37	4.33
CAn 17	12.65	17	2.15
US 28	11.87	28	3.32
US 15	12.65	15	1.90
US 10	12.8	10	1.28
N-pHURIC 10/55	12.8	10	1.28
N-pHURIC 15/49	12.7	15	1.91
N-pHURIC 28/27	11.8	28	3.30

Maintaining a Micro-irrigation System

Routine Maintenance

By Larry Schwankl, UC Irrigation Specialist, and
Terry Prichard, UC Water Management Specialist

Micro-irrigation systems are often automated and therefore typically require little attention for irrigation purposes. Nonetheless, they may require a significant amount of maintenance to continue operating at maximum efficiency.

Routine maintenance can include checking for leaks, backwashing filters, periodically flushing lines, chlorinating, acidifying, and cleaning or replacing clogged emitters.

Cleaning Filters

Filters — whether screen or media — should be backwashed periodically to clear any collected particulate or organic matter. Clogged filters can reduce pressure to the system, lowering the water application rate. Backwashing can be done either manually or automatically. Depending on the design of the screen filter, manual backwashing is accomplished either by physically removing and cleaning the screen or by opening a valve to allow water pressure to scrub the screen clean. Manually backwashing the media filter requires initiating a backwash cycle in which water is circulated from bottom to top, causing the media to be suspended and agitated, which washes the particulate matter out of the filter media.

Automatic backwashing of screen or media filters accomplishes the same task on an automatic, periodic basis. Most automatic backwash systems have an overriding pressure-sensing system that will initiate backwashing if a preset pressure differential across the filter — is exceeded.

Flushing Lines

Also, the mainlines, submains, and particularly the lateral lines should be flushed periodically to clear away any accumulated particulates. Mainlines and submains are flushed by opening the flush valves built into the system for that purpose. When the system is designed, the flush valves should be made large enough to allow for the water velocity to move particulates out of the system.

Lateral lines are flushed by opening the lines and allowing them to clear. This measure is essential, since the filters trap only the large contaminants entering the system, causing lateral lines to collect material that may eventually clog the emitters. Flushing clears the system of many contaminants.

How often the system should be flushed depends on the irrigation water quality and on the degree of filtration. Generally, flushing should be performed bi-weekly, although less-frequent flushing may be found to be adequate. The laterals should also be flushed following fertilizer or chemical injection and any periodic chlorine injection. A simple way to determine how often flushing is needed is to watch to see how much foreign material is removed during flushing. If very little foreign material is flushed out, especially from the lateral lines, flushing can probably be performed less often. The reverse also holds true — if large amounts of material wash out during flushing, flushing should take place more often.

Automatic flushing valves, installed on the ends of the laterals, are also available. These valves remain open until enough pressure builds up in the system to close them. Automatic flushing valves work well in some systems, but under some conditions do not provide a long enough or vigorous enough flushing period to rid the lateral lines of all contaminants.

Chlorination

Waters with a high organic load (algae, moss, bacterial slimes) should undergo chlorination with chlorine gas, sodium hypochlorite, or calcium hypo-chlorite. Whether chlorination should take place continually (1 to 2 ppm free chlorine at the end of lateral line) or periodically (approximately 10 ppm free chlorine at lateral end) depends on the severity of the clogging. Continual chlorination is usually necessary where the clogging potential is severe. Surface water sources are more likely than groundwater sources to cause organic clog-ging. Well water pumped into and stored in a pond or reservoir should be considered a surface water source. *(See "Chlorination," page 93 for more de-tailed information.)*

Acidification

Acidification may be required for irrigation water having a tendency to form chemical precipitates (e.g., lime or iron). Groundwater sources are most susceptible to chemical precipitation. *(See "Chemical Precipitate Clogging," page 103 of this handbook.)*

Acidification to lower the pH of the water to 7.0 or below will usually be sufficient to minimize chemical precipitate problems. Acids that can be added to the irrigation water include sulfuric, hydrochloric, or phosphoric acid. A nitro-gen fertilizer/sulfuric acid mix is frequently used and is safer to handle. Acidification has the added benefit of increasing the efficacy of chlorine addi-tions.

Emitter Maintenance

Emitters may have to be cleaned or replaced because of clogging. The laterals and emitters should be inspected routinely to identify drip emitters that are completely clogged, although this will probably not identify emitters in which the flow has been only reduced. Partially clogged emitters will be located only if water is collected from the emitters to determine their discharge rate.

Micro-sprinklers are easier than drip emitters to inspect visually. Even partial clogging of micro-sprinklers is often visually evident.

The first step in cleaning emitters is determining what is causing the clogging. Emitters and material caught when the laterals are flushed can be examined to determine the cause. If organic matter is the culprit, a high level of chlorine (approximately 50 ppm) can be injected into the line for a number of hours and then allowed to sit for about twenty-four hours. The lines should then be thoroughly flushed. If chemical precipitation is causing the clogging, acid can be injected for a number of hours to lower the pH to approximately 5.0. The acid should be allowed to sit in the line for twenty-four hours and then flushed. If this does not clear the emitters, they may have to be replaced, in which case it is usually wise to leave the clogged emitter in the line and to simply install a new emitter nearby. Although some brands of drip emitters can be disassembled and cleaned, nearly all are permanently sealed. Most micro-sprinklers clog at the orifice in the head and can be cleaned and reinstalled.

If pressure-compensating emitters are being used, the manufacturer should be contacted to determine the minimum pH allowable. The flexible orifice in some pressure-compensating emitters will be damaged by water with a pH of 4.0 or below.

Less-Frequent Tasks

Other maintenance tasks to be carried out on a less-frequent basis include inspecting the filter media, inspecting the pressure-regulating valve, and replacing pressure gauges.

Filter media tend to cake together over time and as a result may fail to provide good filtration. Frequent backwashing may be symptomatic of such a problem. Sand media should be replaced if this occurs. When the old media is removed, the underdrain system should be inspected. Even if the sand media appears to be in good condition, additional media may be added periodically, since some of the sand is invariably lost during the backwash cycle.

Adjustable pressure-regulating valves, set at installation, should be inspected and adjusted periodically to see that the correct operating pressure is maintained. Pre-set pressure-regulators should be inspected to ensure that they are operating properly. Foreign material in the line may jam the adjustment mechanism and inhibit operation.

Pressure gauges tend to eventually wear out and should be replaced if the accuracy is in question. Liquid-filled pressure gauges, which are slightly more expensive, may be a good replacement choice. Gauges must be scaled to operate in a pressure range appropriate to the system.

Leaks

Micro-irrigation systems should be inspected regularly for leaks. This task can be performed when the system is checked for clogged emitters. Leaks can occur in hardware — compression fittings, end closures, emitter barbs, micro-sprinklers, and hose adapters — or when the above-ground polyethylene tubing is damaged by farm equipment, harvest activity, or animals.

Chlorination

*By Larry Schwankl, UC Irrigation Specialist, and
Terry Prichard, UC Water Management Specialist*

Chlorine is often added to irrigation water to oxidize and destroy biological microorganisms such as algae and bacterial slimes. While these microorganisms may be present in water from any source, they are most likely to be present at high levels in surface water from rivers, canals, reservoirs, and ponds.

When water containing high levels of microorganisms is introduced into a micro-irrigation system, emitters can become clogged. Using good filters (such as media filters) and acidifying the water can cut down on organic clogging, but the best way to deal with the problem is to add a biocide such as chlorine.

Dissolving chlorine in water produces hypochlorous acid, which becomes ionized, forming an equilibrium between the hypochlorous acid and hypochlorite — referred to collectively as the *free available chlorine.* Hypochlorous acid is a more powerful biocide than hypochlorite. Acidifying the water tends to favor the production of hypochlorous acid and thus makes the chlorine added more effective. It is important to NOT mix chlorine and acids together, since this causes the formation of chlorine gas, which is toxic.

Sources of Chlorine

The most common sources of chlorine are sodium hypochlorite (a liquid), calcium hypochlorite (powder or granules), and chlorine gas.

Sodium hypochlorite is usually available with up to 15% available chlorine. Household bleach is sodium hypochlorite, with 5.25% active chlorine. To determine the chlorine injection rate when using sodium hypochlorite, the following formula can be used:

$$\text{Chlorine injection rate (gal/hour)} = \text{System flow rate (gpm)} \times \text{Desired chlorine concentration (ppm)} \times 0.006 \div \text{Strength of chlorine solution (\%)} \quad (1)$$

Example: Determine the appropriate injection rate of household bleach (5.25% active chlorine) to obtain a 5 ppm chlorine level in the irrigation system water. The irrigation system flow rate is 100 gpm.

Chlorine injection = 100 gpm x 5 ppm x 0.006 ÷ 5.25% = 0.57 gal/hr

Calcium hypochlorite with 65-70% available chlorine can usually be obtained. In using the formula given above, note that 12.8 pounds of calcium hypochlorite added to 100 gallons of water will form a 1% chlorine solution. A 2% chlorine solution would, therefore, require adding 25.6 pounds of calcium hypochlorite to 100 gallons of water. Any chlorine stock solution can be mixed following the same pattern.

Chlorine gas contains 100% available chlorine. While using chlorine gas is generally considered the least expensive method of injecting chlorine, it is also the most hazardous and requires extensive safety precautions. The chlorine gas injection rate can be determined from the following formula:

$$\text{Chlorine gas injection rate (lbs/day)} = \text{System flow rate (gpm)} \times \text{Desired chlorine concentration (ppm)} \times 0.012 \qquad (2)$$

Desirable Chlorine Injection Rates

If the irrigation water has high levels of algae and bacteria, continuous chlorination may be necessary. The recommended level of free available chlorine is 1 to 2 ppm, measured at the end of the farthest lateral with a good quality pool/spa chlorine test kit.

Periodic injection (once every two to three weeks), at a higher chlorine rate (10-20 ppm) may be appropriate where algae and bacterial slimes are less of a problem. How frequently chlorine injection should be performed depends on the extent of organic clogging.

Superchlorination — bringing chlorine concentrations to within 500 to 1000 ppm — is recommended for reclaiming micro-irrigation systems clogged by algae and bacterial slimes. Superchlorination requires special care to avoid damage to plants and to irrigation components.

Precautions

Following are precautions to be followed in performing chlorination:

• Inject the chlorine upstream from the filter to help keep the filter clean and so that the filter can remove any precipitates that may be caused by the chlorine injection. Chlorine is a very effective oxidizing agent and will cause any iron and manganese present in the water to precipitate and clog the emitters.

• Chlorine compounds should be stored separately in fiberglass or epoxy-coated plastic tanks. Acids and chlorine should never be stored together.

• Do not inject chlorine when fertilizers, herbicides, and insecticides are being injected, since the chlorine may destroy the effectiveness of these compounds.

• When mixing stock chlorine solutions, always add the chlorine source (dry or liquid) to the water, not vice versa.

Heading Off Problems

Assessing Water Quality
By Blaine Hanson, UC Irrigation and Drainage Specialist

The irrigation water to be used in a drip system should be carefully evaluated to assess any potential clogging problems. Materials such as sand, silt, and algae suspended in the water can block emitter flow passages or settle out in the drip lines wherever water velocity is low. Constituents such as calcium, bicarbonate, iron, manganese, and sulfide can also precipitate to clog emitter flow passages, and where iron and manganese concentrations are high enough, iron slimes and bacteria can grow, clogging drip lines.

Table 1 provides criteria developed from numerous evaluations of the effect of water quality on emitter flowrate. This information can be used to assess irrigation water for clogging potential.

Table 1. Relative clogging potential of irrigation water in micro-irrigation systems.

Water characteristics	*minor*	*moderate*	*severe*
Maximum suspended solids (ppm)	<50*	50-100	>100
pH	<7.0	7.0 - 8.0	>8.0
Maximum total dissolved solids (ppm)	<500	500-2000	>2000
Maximum manganese concentration (ppm)	<0.1	0.1 - 1.5	>1.5
Maximum iron concentration (ppm)	<0.2	0.2 - 1.5	>1.5
Maximum hydrogen sulfide concentration (ppm)	<0.2	0.2 - 2.0	>2.0
Bacterial population (maximum number per ml)	<10,000	10,000 - 50,000	>50,000

* < = less than. > = more than

Sources: D.A. Bucks, F. S. Nakayama, and R. G. Gilbert. 1979. "Trickle irrigation water quality and preventive maintenance." *Agricultural Water Management,* Vol. 2:149-62; F.S. Nakayama and D.A. Bucks. 1991. "Water quality in drip/trickle irrigation: A Review." *Irrigation Science,* Vol. 12:187-92.

Notes

1. Bicarbonate concentrations exceeding about 2 meq/liter and pH exceeding about 7.5 can cause calcium carbonate precipitation.
2. Calcium concentrations exceeding 2-3 meq/liter can cause precipitates to form during injection of some phosphate fertilizers.
3. High concentrations of sulfide ions can cause iron and manganese precipitation. Iron and manganese sulfides are very insoluble, even in acid solutions.

Chemical Constituents

Irrigation water should be analyzed for the following:

1. electrical conductivity (EC)—a measure of the total dissolved salts (TDS). An approximate equation relating TDS to EC is:

$$TDS(ppm) = 640 \times EC \ (dS/m \ or \ mmhos/cm)$$

2. pH
3. calcium (Ca)
4. magnesium (Mg)
5. sodium (Na)
6. chloride (Cl)
7. sulfate (SO_4)
8. carbonate/bicarbonate (CO_3 / HCO_3)
9. iron (Fe)
10. manganese (Mn)

Units of Measurement

The most common measurement unit for reporting concentrations is parts per million (ppm). Concentrations are also reported as milligrams per liter (mg/*l*). For practical purposes, ppm equals mg/*l* for irrigation water.

Concentrations may be reported in kilograms per cubic meters (kg/m³) which is the SI unit. Kg/m³ is the same as mg/*l*.

Concentrations may also be reported in milliequivalents per liter (meq/*l*). Conversion factors are needed to convert from mg/*l* to meq/*l* and vice versa. Table 2 list conversion factors for common constituents.

Grains per gallon may be used as a concentration unit. To convert grains per gallon to mg/*l*, multiply the grains per gallon by 17.12.

Table 2. Conversion factors: parts per million and milliequivalents per liter.

constituent	convert ppm to meq/*l*	convert meq/*l* to ppm
	multiply by	
Na (sodium)	0.043	23
Ca (calcium)	0.050	20
Mg (magnesium)	0.083	12
Cl (chloride)	0.029	35
SO_4 (sulfate)	0.021	48
CO_3 (carbonate)	0.033	30
HCO_3 (bicarbonate)	0.016	61

Examples:

  1. convert 415 ppm of Na to meq/*l*:

     meq/*l* = 0.043 x 415 ppm = 17.8

  2. convert 10 meq/*l* of SO$_4$ to ppm:

     ppm = 48 x 10 meq/*l* = 480

The quality of the data should be evaluated using the following procedures:

a. The sum of the cations (Ca, Mg, Na), expressed in milliequivalents per liter (meg/l) should about equal the sum of the anions (Cl, CO_3, HCO_3, SO_4). If the sums are exactly equal, then one of the constituents was found by differences.

b. The sum of the cations and the sum of the anions should each equal about 10 times the EC.

If these procedures reveal poor quality, the chemical analysis should be repeated.

Evaluating Your Water Quality

The following steps are guidelines for evaluating your water quality. Refer to Table 1 to assist you.

1. What is the total dissolved solids concentration? If the electrical conductivity is given only, multiply this EC (mmhos/cm) by 640 to determine the total dissolved solids.
2. What is the calcium concentration? If the calcium concentration exceeds 2-3 meq/*l*, read the chapter "Chemical Precipitate Clogging".
3. What is the bicarbonate concentration? If the bicarbonate concentration exceeds about 2 meq/*l*, read the chapter "Chemical Precipitate Clogging".
4. What is the iron and manganese concentrations? If either concentration exceeds about 0.2 ppm, read the chapter "Chemical Precipitate Clogging".

Hardness and Alkalinity

The hardness and alkalinity of a water may be reported for a water analysis, although these characteristics normally are not used for assessing potential clogging problems in drip irrigation.

The hardness of the water is primarily due to calcium and magnesium ions. Hard water will tend to precipitate calcium carbonate. Thus, the higher the hardness, expressed in terms of calcium carbonate, the higher the potential for calcium carbonate precipitation in drip irrigation systems. Classifications of hardness are:

0-75 mg/*l* - soft
75-150 mg/*l* - moderately hard
150-300 mg/*l* - hard
more than 300 mg/*l* - very hard

Alkalinity of a water is a measure of its ability to neutralize acids. Alkalinity is caused mostly by carbonate and bicarbonate ions. Lowering the pH of waters with a large alkalinity will require more acid than for waters with a lower alkalinity.

Table 2 gives water quality data from the analysis of two irrigation water samples. *Examples 1* and *2* use the water quality data from *Table 2* to evaluate the clogging potential of these irrigation waters.

Table 2. Water quality analysis of two irrigation water samples.

Water 1	Water 2
EC = 2.51 dS/m (1900 ppm)[1]	*EC = 0.87 dS/m (560 ppm)[2]*
pH = 7.4	*pH = 7.7*
Ca = 13.3 meq/l	*Ca = 1.9 meq/l*
Mg = 10.1 meq/l	*Mg = 1.3 meq/l*
Na = 5.4 meq/l	*Na = 5.5 meq/l*
Cl = 4.5 meq/l	*Cl = 2.0 meq/l*
HCO$_3$ = 5.2 meq/l	*HCO$_3$ = 2.0 meq/l*
SO$_4$ = 19 meq/l	*SO$_4$ = 4.7 meq/l*
Mn= less than 0.1 ppm	*Mn= 2.6 ppm*
Fe = less than 0.1 ppm	*Fe = 0.65 ppm*

[1] Total dissolved salts = 757 x EC
[2] Total dissolved salts = 644 x EC

Example

Example 1. The relatively high total dissolved salts (TDS) (1900 ppm) indicates that *Water 1* has some clogging potential. This is verified by the relatively high bicarbonate concentration (5.2 meq/l), compared to the standard of 2.0 meq/l. The calcium concentration and the bicarbonate concentration together suggest that calcium carbonate could clog the emitters, particularly if the pH were to rise as a result of any chemical injection. The iron and manganese concentrations indicate little potential for clogging from precipitation of those elements.

Example 2. The analysis of *Water 2* reveals little potential for clogging from total dissolved salts (560 ppm), but the pH and bicarbonate concentrations indicate that clogging might result from calcium carbonate precipitation. The levels of manganese and iron indicate a severe potential for clogging from manganese oxide precipitation and iron oxide precipitation.

References

Nordell, E. 1961. Water Treatment for Industrial and Other Uses. Reinhold Publishing Corporation, N.Y. 598 p.

Barnes, D. and F. Wilson. 1983. Chemistry and Unit Operations in Water Treatment. Applied Science Publishers, London and N.Y. 325 p.

Sawyer, C.N. and P.L. McCarty. 1967. Chemistry for Sanitary Engineers. McGraw-Hill Book Company, N.Y. 518 p.

Chemical Precipitate Clogging

By Larry Schwankl, UC Irrigation Specialist,
Blaine Hanson, UC Irrigation and Drainage Specialist, and
Terry Prichard, UC Water Management Specialist

Precipitating chemicals and organic contaminants can clog micro-irrigation systems. When a micro-irrigation system using groundwater for irrigation becomes clogged, the usual cause is chemical precipitation from calcium carbonate (lime), iron, or manganese in the irrigation water.

Lime Precipitation

Calcium carbonate (lime) precipitation is the most common cause of chemical clogging in micro-irrigation. Water with a pH of 7.5 or above and bicarbonate levels of 2 meq/l (120 ppm) is susceptible to lime precipitation if comparable levels of calcium are present naturally in the system or if a compound containing calcium is injected into the system.

The usual treatment for lime precipitation is to acidify the water to lower the pH to 7.0 or below. Litmus paper, colormetric kits, or portable pH meters can be used to determine the water's pH. Sulfuric acid is usually used to reduce pH, but phosphoric acid and hydrochloric acid may also be used. Since handling acids is hazardous, some water managers prefer to use one of the safer acid/fertilizer compounds now available. Researchers are evaluating other compounds — including a phosonate material and several polymer materials — to determine their efficacy in preventing calcium carbonate precipitation.

Iron and Manganese

Iron and manganese precipitation can cause clogging even at low concentrations: iron: 0.3 ppm or greater, manganese: 0.15 ppm or greater. These compounds, which are most often present in groundwater, are in a soluble reduced state in the well, but oxidize and precipitate as very small but solid particles when exposed to the atmosphere. Iron and manganese will precipitate across a wide range of pH (iron, for example, will precipitate at pH 4.0-9.5), which include the pH levels of almost all naturally occurring waters.

Iron precipitate is characterized by a reddish stain and rust particles in the water. Manganese precipitate has a similar appearance, but the stain is darker — nearly black in color.

Iron/manganese precipitation is further complicated by bacteria that use iron/manganese as energy sources. These bacteria form filamentous slimes that can clog filters and emitters and that can also provide the matrix or glue to hold other contaminants in the system. Iron bacteria can be controlled by injecting chlorine continually at 1-2 ppm residual (at the end of the line) or intermittently at 10-20 ppm residual.

How to Mitigate Chemical Iron or Manganese Precipitation

The following measures can be taken to mitigate chemical iron or manganese precipitation:

1. Aeration and settling. Water can be pumped into a pond or reservoir and allowed to aerate from contact with the atmosphere. The iron precipitate is then allowed to settle out. Additional aeration of the water may be necessary to ensure that the iron is oxidized. After the iron settles, the water can be drawn off for use.

2. Chlorine precipitation and filtration. Injecting chlorine into the water will oxidize the dissolved (ferrous) iron, causing it to precipitate. The precipitated iron (ferric oxide) can then be filtered out, preferably with a sand media filter, which can be readily backwashed.

3. pH control. Where the potential for iron precipitation exists, lowering the pH in the system to less than 4.0 will keep the iron from precipitating. The cost of this practice may limit its use.

4. Chelation. In municipal water treatment, a polyphosphate, such as sodium hexametaphosphate, is added to the water before the iron is oxidized. This prevents agglomeration of the small individual particles. Recommended injection rates are 2 mg/l of sodium hexametaphosphate for each 1 mg/l of iron or manganese. Since this practice is expensive it should only be used in agricultural systems after careful evaluation.

Miscellaneous Compounds

Other compounds that can cause clogging include magnesium carbonate, calcium sulfate, and zinc injected in sulfate form. Adding anhydrous or aqua ammonia to irrigation water will increase its pH, possibly facilitating the precipitation of calcium or magnesium compounds. Adding phosphate fertilizers may also cause the phosphate to react with calcium or magnesium, resulting in a precipitate. This can be prevented by adding acid to significantly lower the pH of the water.

Table 1 summarizes recommended treatments for various types of chemical and biological clogging.

Table 1. Water treatments to prevent clogging in micro-irrigation systems

Problem	Treatment Options

Carbonate precipitation (white precipitate)
    HCO_3 greater than 2.0 meq/l
    pH greater than 7.5
	1. Continuous injection: maintain pH between 5 and 7
	2. Slug injection: maintain pH at under 4 for 60-90 minutes daily

Iron precipitation (reddish precipitate)
    Iron concentrations greater than 0.1ppm
	1. Aeration and settling to oxidize iron. (Best treatment for high concentrations—10 ppm or more.)
	2. Chlorine precipitation—injecting chlorine to precipitate iron:
	        a. use an injection rate of 1 ppm of chlorine per 0.7 ppm of iron
	        b. inject in front of the filter so that the precipitate is filtered out

Manganese precipitation (black precipitate)
    Manganese concentrations greater
    than 0.1 ppm
	1. Inject 1.3 ppm of chlorine per 1 ppm of manganese in front of the filter

Iron bacteria (reddish slime)
    Iron concentrations greater
    than 0.1 ppm
	1. Inject chlorine at a rate of 1 ppm free chlorine continuously or 10 to 20 ppm for 60 to 90 minutes daily.

Sulfur bacteria (white cotton-like slime)
    sulfide concentrations greater
    than 0.1 ppm
	1. Inject chlorine continuously at a rate of 1 ppm per 4 to 8 ppm of hydrogen sulfide, or
	2. Inject chlorine intermittently at 1 ppm free available chlorine for 60 to 90 minutes daily.

Algae, slime	1. Inject chlorine at a rate of 0.5 to 1 ppm continuously or 20 ppm for 60 minutes at the end of each irrigation cycle.

Iron sulfide (black sand-like material)
    Iron and sulfide concentrations
    greater than 0.1 ppm
	1. Dissolve iron by injecting acid continuously to lower pH to between 5 and 7.

Root Intrusion

By Larry Schwankl, UC Irrigation Specialist, and
Terry Prichard, UC Water Management Specialist

Roots can clog subsurface drip irrigation systems by intruding into emission orifices. Small roots tend to proliferate near the drip tape or tubing where water is present and have been found to grow through the emitter pathway into the drip tape/tubing. The result is often partial or total plugging of the drip emitter pathway, drastically reducing emitter flowrates.

Permanent Crops

Because subsurface drip systems installed in permanent crops such as trees and vines are left in place for a number of years, they are more likely to be invaded by roots than are systems installed in annual crops. On the other hand, the longer the subsurface drip system remains operative, the lower the annualized cost of the system.

Root intrusion is magnified in permanent crops since roots do not die off as they do in annual crops. Root growth may even take place in permanent crops during the off season, with a flush of root growth occurring in the late fall and early spring even though the plant canopy may be dormant and the irrigation system shut down.

Various water management strategies and chemical treatments can be used to combat root intrusion. Irrigating daily to keep the area surrounding the subsurface tape/tubing saturated seems to create an environment hostile to root growth. This may explain why root intrusion is more prevalent late in the season when the irrigation system is shut down, but while just enough water remains inside the tape/tubing to attract roots.

Inexpensive chemical treatments can be used to kill and remove root residue by oxidation. Periodic injections of chlorine or acid products can destroy intruding root hairs, although the use of oxidation materials may be less effective on woody perennials such as trees and vines. Weekly chlorine injections resulting in 10 ppm free chlorine concentrations have shown promise in controlling root intrusion into subsurface drip systems in walnuts. Injecting acid (sulfuric, hydrochloric, or phosphoric) to lower the pH of the irrigation water may also be effective, but little research has been done to verify this approach. Caution is advised in reducing pH levels below 4.0, since emitter damage may result. Flushing lateral lines frequently with water, along with using chlorine or acid treatments to remove any root matter from the system, is also strongly recommended.

One drip tubing product presently on the market is designed to prevent root intrusion. This product relies on in-line emitters impregnated with the chemical compound Trifluralin, which is released gradually to inhibit root growth once the system is in place.

Chemical treatments and flushing are also good techniques for preventing clogging caused by bacterial slimes and particulate matter (silts and clays).

Injecting copper sulfate may also be effective, but this chemical treatment is still being investigated. It has been postulated that since copper —which is toxic to roots — is quickly absorbed and is held on the soil particles surrounding the drip tape/tubing, it may form a more permanent barrier to root intrusion.

Detecting
Root Intrusion

It is important to monitor a subsurface drip irrigation system to determine if emitter clogging is occurring. Since emitters are buried, making discharge difficult to measure, monitoring flowrate and pressure in the system —preferably in small blocks — is recommended. Clogged emitters discharge less water (measured at the same pressure) than unclogged emitters. This reduced discharge can be detected with a flowmeter. If emitter clogging does appear to be occurring, additional investigations should be carried out to determine whether the clogging is being caused by root intrusion, chemical precipitation, biological growth, or particulate matter.

Salt Patterns Under Drip Irrigation
By Blaine Hanson, UC Irrigation and Drainage Specialist

Salt movement is governed by water movement. Under drip irrigation, water moves in a more or less radial pattern around the emitter, and salinity in the soil eventually mirrors this pattern.

Figure 1 shows salinity patterns for two leaching fractions of 5% and 25% under surface drip irrigation. (The leaching fraction is applied water that exceeds the evapotranspiration since the last irrigation and that therefore percolates down past the root zone as excess water.) The following can be concluded from these patterns:

Salinity Patterns After Irrigation

• Salinity is lowest directly beneath the plant row and emitter. This low salinity zone is greatest for the high leaching fraction and lowest for the low leaching fraction.

• Salinity gradually increases as the distance from the emitter increases. The increase is smallest in the vertical direction and largest in the horizontal direction. With the low leaching fraction, levels of increased salinity occur closer to the emitter.

• Salinity is highest midway between emitters. This zone is smallest for the high leaching fraction and largest for the low leaching fraction. At the midway point, salinity decreases as the depth increases.

These salt patterns reflect water movement during and between irrigations. During irrigations, salt leaching occurs in the vicinity of the emitter. The infiltrating water carries these leached salts away from the emitter. As the horizontal distance from the emitter increases, soil salinity increases because the amount of leaching decreases. Salt accumulation is highest midway between emitters because no leaching occurs in those areas.

When rainfall or water applied with another irrigation system infiltrates the soil, salts accumulated near the surface are carried downward with the infiltrating water. Leaching can therefore occur in areas with high salt accumulations, thereby lowering the salinity near the surface. The leaching process moves salt downward as a front or slug as shown in *Figure 1. Figure 1* also shows salt distribution after rainfall.

For each leaching fraction, the initial zone of high salinity midway between emitters moves downward in the soil profile as a slug of salt. The greater the rainfall, the deeper the salt slug will move.

Effect on Plant Growth

Concerns have been raised about the effect on crop growth of this salt front moving downward. *Figure 2* shows a distribution of soil salinity and root density. Again, soil salinity is highest near the surface. If rainfall occurs, these salts will move downward, but most of the plant roots are located near the emitter, where salinity is lowest, with substantially fewer roots growing in areas of maximum salinity and in the soil profile immediately below those areas. During a rainfall, therefore, salts near the soil surface will be leached downward into areas with relatively low root densities. This suggests that the salt movement will have little effect on crop growth.

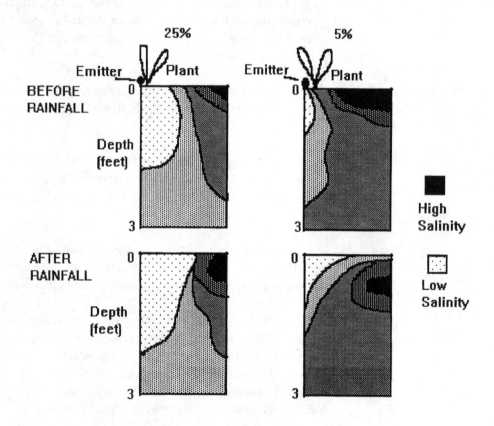

Figure 1. Salt patterns under drip irrigation.

In *Figure 1,* the emitter and plant are at the same location. Other data on salinity distributions under drip irrigation show that as the emitter is moved away from the plant, the high salinity zone can shift toward the plant, and in some cases can occur at the plant within the active root zone.

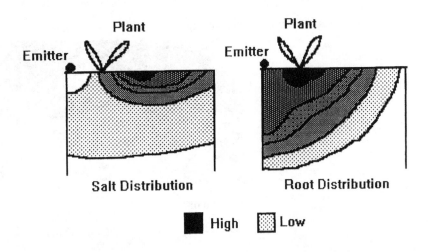

Figure 2. Distribution of soil salinity and root density.

Subsurface Drip Irrigation

The salt patterns that form under buried drip irrigation are different from those that form under surface drip irrigation. *Figure 3*, which illustrates a salt pattern occurring under subsurface drip irrigation, shows very high soil salinity levels near the ground surface and extending through the top few inches of the soil surface. These salinity levels can exceed 10 dS/m, depending on the salinity of the irrigation water. Salinity decreases with depth through the soil profile and increases with horizontal distance from the emitter. Near the drip tape, soil salinity is relatively low, and directly beneath the drip tape, soil salinity changes only slightly with depth.

This salt pattern shows that drip irrigation does not cause salt leaching above the drip tape, but does cause substantial leaching beneath the tape and in its immediate vicinity. Leaching diminishes under the tape as horizontal distance increases.

Leaching

Since drip irrigation does not accomplish leaching above the drip tape, leaching will have to be done with another irrigation system or through rainfall. If rainfall is used for leaching, the drip system should be operated to replenish the soil water content to field capacity and to increase the leaching effectiveness of the rainfall, since no leaching will take place until the soil moisture exceeds field capacity. *Figure 4* shows the salt pattern after about 14 inches of rainfall.

To lessen salinity just before planting, the bed of the furrow can be built-up and the drip system operated to accumulate the salt in the built-up portion of the bed. The top of the bed can then be removed before planting to leave a relatively low-salt seedbed.

Subsurface drip irrigation can present a salinity hazard if rainfall during the crop season moves a slug of very high salt concentration down into the root zone, to the detriment of shallow-rooted crops. To minimize salinity damage, the drip system can be operated during the rainfall to dilute the salt.

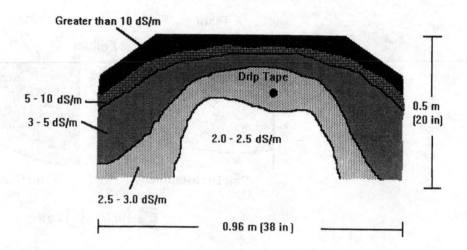

Figure 3. Salt pattern under subsurface drip irrigation.

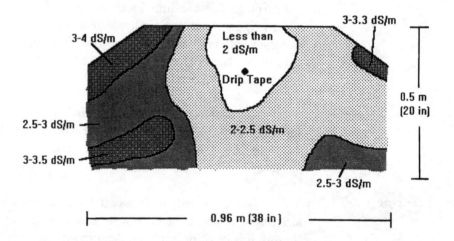

Figure 4. Salt pattern after 14 inches of rainfall.

Leaching Fractions Under Drip Irrigation

Leaching fractions function differently under drip irrigation than in traditional irrigation practices. With conventional methods, leaching fractions result in an average soil salinity. But under drip irrigation, where leaching ranges from very small amounts midway between laterals to very large amounts beneath the emitter, the leaching fraction results in an area of low salinity soil adjacent to the emitter. The greater the leaching fraction, the greater the volume of low salinity soil.

How much leaching is necessary with drip irrigation? Table 1 list some leaching requirements for permanent crops for irrigation water salinity of 0.5 and 1 dS/m. These requirements are based on research by the U.S. Salinity Laboratory.

The amount of applied water needed to satisfy both ET and leaching can be calculated by:

$$\text{Applied water} = \frac{\text{Crop Evapotranspiration}}{(1-\text{Leaching Requirement})}$$

where the leaching requirement is expressed as a decimal fraction.

Table 1. Recommended leaching requirements for some permanent crops.

Crop	Leaching Requirement (%)	
	0.5 dS/m	*1 dS/m*
Almond	*8*	*20*
Apricot	*7*	*19*
Grape	*8*	*20*
Orange	*6*	*18*
Peach	*6*	*18*

References

Hoffman G.J., S.L. Rawlins, J.D. Oster, and S.D. Merrill. 1979. "Leaching requirement for salinity control I. Wheat, sorghum, and lettuce. *Agricultural Water Management,* Vol. 2: 177-92.

Hoffman, G.J., and J.A. Jobes. 1983. "Leaching requirement for salinity control III. Barley, cowpea, and celery. *Agricultural Water Management,* Vol. 6: 1-14.

Hoffman, G. J. , M. C. Shannon, and J.A. Jobes. 1985. "Influence of rain on soil salinity and lettuce yield." In *Irrigation in Action,* Proceedings of the Third International Drip/ Trickle Irrigation Congress, November 18-21, 1985, Fresno, California.

Jobes, J.A., G. J. Hoffman, and J.D. Wood. 1981. "Leaching requirement for salinity control II. Oat, tomato, and cauliflower. *Agricultural Water Management,* Vol. 4: 393-407.

Moshrefi, N. and F. Beese. 1985. "Effect of irrigation system on salt and root distribution." In *Irrigation In Action,"* Proceedings of the Third International Drip/Trickle Irrigation Congress, November 18-21, 1985, Fresno, California.

Plaut, Z., M. Rom, and A. Meiri. 1985. "Effect of irrigation system on salt and root distribution." In *Irrigation In Action,* Proceedings of the Third International Drip/ Trickle Irrigation Congress, November 18-21, 1985, Fresno, California.

Yaron, B., J. Shalhevet, and D. Shimshi. 1973. "Patterns of salt distribution under trickle irrigation." In *Physics of Soil Water and Salt,* Ecological Studies 4. Springer-Verlag.

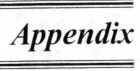

Appendix

Table 1. Pressure loss in 1/2" to 2" Class 160 PVC pipe.

			PRESSURE LOSS TABLES			
		HAZEN-WILLIAMS EQUATION (C = 140) CLASS 160 PVC				
			PRESSURE LOSS (PSI/100 FEET)			
			PIPE SIZE			
FLOW	1/2"	3/4"	1"	1 1/4"	1 1/2"	2"
GPM						
0.1	0.003					
0.2	0.012					
0.3	0.026					
0.4	0.044	0.013				
0.5	0.066	0.019				
0.6	0.093	0.027				
0.7	0.123	0.035				
0.8	0.158	0.045	0.013			
0.9	0.196	0.056	0.017			
1	0.239	0.069	0.020			
1.1	0.285	0.082	0.024			
1.2	0.334	0.096	0.028			
1.3	0.388	0.112	0.033			
1.4	0.445	0.128	0.038			
1.5	0.506	0.145	0.043	0.013		
1.6	0.570	0.164	0.048	0.014		
1.7	0.638	0.183	0.054	0.016		
1.8	0.709	0.204	0.060	0.018		
1.9	0.783	0.225	0.066	0.020		
2	0.861	0.248	0.073	0.022		
2.1	0.943	0.271	0.080	0.024		
2.2	1.028	0.296	0.087	0.026		
2.3	1.116	0.321	0.095	0.028	0.015	
2.4	1.208	0.347	0.102	0.031	0.016	
2.5	1.302	0.374	0.110	0.033	0.017	
2.6	1.400	0.403	0.119	0.035	0.018	
2.7	1.502	0.432	0.127	0.038	0.020	
2.8	1.606	0.462	0.136	0.041	0.021	
2.9	1.714	0.493	0.145	0.043	0.022	
3	1.825	0.525	0.155	0.046	0.024	
3.1	1.940	0.558	0.165	0.049	0.025	
3.2	2.057	0.592	0.174	0.052	0.027	
3.3	2.178	0.626	0.185	0.055	0.028	
3.4	2.302	0.662	0.195	0.058	0.030	
3.5	2.429	0.698	0.206	0.061	0.032	
3.6	2.559	0.736	0.217	0.065	0.033	
3.7	2.692	0.774	0.228	0.068	0.035	

Table 1. Pressure loss in 1/2" to 2" Class 160 PVC pipe (continued)

		PRESSURE LOSS TABLES				
	HAZEN-WILLIAMS EQUATION (C = 140) CLASS 160 PVC					
		PRESSURE LOSS (PSI/100 FEET)				
			PIPE SIZE			
FLOW	1/2"	3/4"	1"	1 1/4"	1 1/2"	2"
GPM						
3.8	2.828	0.813	0.240	0.072	0.037	
3.9	2.967	0.853	0.252	0.075	0.039	
4	3.110	0.894	0.264	0.079	0.041	
4.1	3.255	0.936	0.276	0.082	0.043	
4.2	3.404	0.979	0.289	0.086	0.045	0.015
4.3	3.556	1.022	0.302	0.090	0.047	0.016
4.4	3.710	1.067	0.315	0.094	0.049	0.016
4.5	3.868	1.112	0.328	0.098	0.051	0.017
4.6	4.029	1.158	0.342	0.102	0.053	0.018
4.7	4.192	1.205	0.356	0.106	0.055	0.018
4.8	4.359	1.253	0.370	0.110	0.057	0.019
5.25	5.146	1.480	0.436	0.130	0.067	0.023
5.5	5.609	1.613	0.476	0.142	0.073	0.025
5.75	6.090	1.751	0.517	0.154	0.080	0.027
6	6.590	1.895	0.559	0.167	0.086	0.029
6.25	7.107	2.044	0.603	0.180	0.093	0.031
6.5	7.643	2.198	0.648	0.193	0.100	0.034
6.75	8.196	2.357	0.695	0.207	0.107	0.036
7	8.767	2.521	0.744	0.222	0.115	0.039
7.25	9.356	2.690	0.793	0.237	0.122	0.041
7.5	9.962	2.865	0.845	0.252	0.130	0.044
7.75	10.586	3.044	0.898	0.268	0.139	0.047
8	11.227	3.228	0.952	0.284	0.147	0.050
8.25	11.885	3.418	1.008	0.301	0.156	0.052
8.5	12.561	3.612	1.065	0.318	0.164	0.055
8.75	13.254	3.811	1.124	0.335	0.173	0.058
9	13.964	4.015	1.184	0.353	0.183	0.062
9.25	14.690	4.224	1.246	0.372	0.192	0.065
9.5	15.434	4.438	1.309	0.390	0.202	0.068
9.75	16.195	4.657	1.373	0.410	0.212	0.071
10	16.972	4.880	1.439	0.429	0.222	0.075
10.5			1.575	0.470	0.243	0.082
11			1.717	0.512	0.265	0.089
11.5			1.865	0.556	0.288	0.097
12			2.017	0.602	0.311	0.105
12.5			2.176	0.649	0.336	0.113
13			2.340	0.698	0.361	0.122

Table 1. Pressure loss in 1/2" to 2" Class 160 PVC pipe (continued).

PRESSURE LOSS TABLES						
HAZEN-WILLIAMS EQUATION (C = 140) CLASS 160 PVC						
PRESSURE LOSS (PSI/100 FEET)						
			PIPE SIZE			
FLOW GPM	1/2"	3/4"	1"	1 1/4"	1 1/2"	2"
13.5			2.509	0.748	0.387	0.130
12.5			2.176	0.649	0.336	0.113
13			2.340	0.698	0.361	0.122
13.5			2.509	0.748	0.387	0.130
14			2.684	0.801	0.414	0.140
14.5			2.864	0.854	0.442	0.149
15			3.050	0.910	0.471	0.159
15.5			3.241	0.967	0.500	0.168
16			3.437	1.025	0.530	0.179
16.5			3.639	1.085	0.561	0.189
17			3.846	1.147	0.593	0.200
17.5			4.058	1.210	0.626	0.211
18			4.275	1.275	0.660	0.222
18.5			4.497	1.341	0.694	0.234
19			4.725	1.409	0.729	0.246
19.5				1.479	0.765	0.258
20				1.550	0.802	0.270
20.5				1.622	0.839	0.283
21				1.696	0.878	0.296
21.5				1.772	0.917	0.309
22				1.849	0.957	0.322
22.5				1.927	0.997	0.336
23				2.008	1.039	0.350
23.5				2.089	1.081	0.364
24				2.172	1.124	0.379
24.5				2.257	1.168	0.393
25				2.343	1.212	0.408
26				2.519	1.303	0.439
27				2.702	1.398	0.471
28				2.890	1.495	0.504
29				3.084	1.595	0.538
30				3.284	1.699	0.572
32				3.701	1.915	0.645
34				4.140	2.142	0.722
36				4.603	2.381	0.802
38				5.087	2.632	0.887
40				5.594	2.894	0.975

Table 2. Pressure loss in 2" to 8" Class 160 PVC pipe.

	PRESSURE LOSS TABLES						
	HAZEN-WILLIAMS EQUATION (C = 150) CLASS 160 PVC						
	PRESSURE LOSS (PSI/100 FEET)						
	PIPE SIZE						
FLOW	2"	2 1/2"	3"	4"	5"	6"	8"
GPM							
10	0.066						
15	0.140						
20	0.238	0.094					
25	0.359	0.142					
30	0.504	0.199	0.076				
35	0.670	0.264	0.102				
40	0.858	0.338	0.130				
45	1.067	0.421	0.162				
50	1.297	0.511	0.197				
55	1.548	0.610	0.235				
60	1.819	0.717	0.276	0.081			
65	2.109	0.831	0.320	0.094			
70	2.419	0.954	0.367	0.108			
75	2.749	1.084	0.417	0.123			
80	3.098	1.221	0.470	0.138			
85	3.466	1.366	0.526	0.154			
90	3.853	1.519	0.585	0.172			
95	4.259	1.679	0.646	0.190			
100	4.684	1.846	0.711	0.209			
105	5.127	2.021	0.778	0.228	0.081		
110	5.588	2.203	0.848	0.249	0.089		
115	6.068	2.392	0.921	0.270	0.096		
120	6.565	2.588	0.996	0.293	0.104		
125	7.081	2.791	1.074	0.316	0.112		
130	7.614	3.001	1.155	0.339	0.121		
135	8.165	3.218	1.239	0.364	0.130		
140	8.734	3.443	1.325	0.389	0.139	0.059	
145	9.321	3.674	1.414	0.415	0.148	0.063	
150	9.925	3.912	1.506	0.442	0.157	0.067	
155	10.546	4.157	1.600	0.470	0.167	0.071	
160	11.185	4.409	1.697	0.498	0.177	0.076	
165	11.841	4.667	1.796	0.528	0.188	0.080	
170	12.514	4.932	1.899	0.558	0.199	0.085	0.023
175	13.204	5.204	2.003	0.588	0.210	0.089	0.025
180	13.911	5.483	2.111	0.620	0.221	0.094	0.026
185	14.635	5.769	2.220	0.652	0.232	0.099	0.027
190	15.376	6.061	2.333	0.685	0.244	0.104	0.029

Table 2. Pressure loss in 2" to 8" Class 160 PVC pipe (continued).

PRESSURE LOSS TABLES							
HAZEN-WILLIAMS EQUATION (C = 150) CLASS 160 PVC							
PRESSURE LOSS (PSI/100 FEET)							
			PIPE SIZE				
FLOW GPM	2"	2 1/2"	3"	4"	5"	6"	8"
195	16.134	6.359	2.448	0.719	0.256	0.109	0.030
200	16.909	6.665	2.565	0.753	0.268	0.115	0.032
205	17.700	6.976	2.685	0.789	0.281	0.120	0.033
210	18.508	7.295	2.808	0.825	0.294	0.125	0.035
215	19.332	7.620	2.933	0.861	0.307	0.131	0.036
220		7.951	3.061	0.899	0.320	0.137	0.038
225		8.289	3.191	0.937	0.334	0.143	0.039
230		8.633	3.323	0.976	0.348	0.148	0.041
235		8.984	3.458	1.016	0.362	0.154	0.043
240		9.341	3.596	1.056	0.376	0.161	0.044
245		9.705	3.736	1.097	0.391	0.167	0.046
250		10.075	3.878	1.139	0.406	0.173	0.048
255		10.451	4.023	1.182	0.421	0.180	0.050
260		10.834	4.170	1.225	0.436	0.186	0.052
265		11.223	4.320	1.269	0.452	0.193	0.053
270		11.618	4.472	1.313	0.468	0.200	0.055
275		12.020	4.627	1.359	0.484	0.207	0.057
280		12.428	4.784	1.405	0.500	0.214	0.059
285		12.842	4.943	1.452	0.517	0.221	0.061
290		13.262	5.105	1.499	0.534	0.228	0.063
295		13.689	5.269	1.548	0.551	0.235	0.065
300		14.122	5.436	1.596	0.569	0.243	0.067
305		14.561	5.605	1.646	0.586	0.250	0.069
310		15.006	5.776	1.696	0.604	0.258	0.071
315		15.457	5.950	1.747	0.622	0.266	0.074
320		15.915	6.126	1.799	0.641	0.274	0.076
325		16.378	6.304	1.852	0.659	0.282	0.078
330		16.848	6.485	1.905	0.678	0.290	0.080
335		17.324	6.668	1.958	0.697	0.298	0.082
340		17.806	6.854	2.013	0.717	0.306	0.085
345			7.042	2.068	0.736	0.315	0.087
350			7.232	2.124	0.756	0.323	0.089
360			7.619	2.238	0.797	0.340	0.094
370			8.016	2.354	0.838	0.358	0.099
380			8.422	2.473	0.881	0.376	0.104
390			8.837	2.595	0.924	0.395	0.109
400			9.261	2.720	0.969	0.414	0.114

Table 2. Pressure loss in 2" to 8" Class 160 PVC pipe (continued).

			PRESSURE LOSS TABLES				
		HAZEN-WILLIAMS EQUATION (C = 150) CLASS 160 PVC					
		PRESSURE LOSS (PSI/100 FEET)					
			PIPE SIZE				
FLOW	2"	2 1/2"	3"	4"	5"	6"	8"
GPM							
410			9.694	2.847	1.014	0.433	0.120
420			10.137	2.977	1.060	0.453	0.125
430			10.588	3.110	1.107	0.473	0.131
440			11.049	3.245	1.156	0.494	0.137
450			11.518	3.383	1.205	0.515	0.142
460			11.997	3.523	1.255	0.536	0.148
470			12.484	3.666	1.306	0.558	0.154
480			12.981	3.812	1.358	0.580	0.160
490			13.486	3.961	1.410	0.602	0.167
500			14.000	4.112	1.464	0.625	0.173
510			14.523	4.265	1.519	0.649	0.180
520			15.055	4.421	1.575	0.673	0.186
530			15.595	4.580	1.631	0.697	0.193
540			16.145	4.742	1.689	0.721	0.200
550			16.703	4.905	1.747	0.746	0.206
560			17.270	5.072	1.806	0.772	0.213
570			17.845	5.241	1.866	0.797	0.221
580			18.429	5.412	1.927	0.823	0.228
590			19.022	5.587	1.989	0.850	0.235
600			19.623	5.763	2.052	0.877	0.243
610			20.233	5.942	2.116	0.904	0.250
620				6.124	2.181	0.932	0.258
630				6.308	2.246	0.960	0.266
640				6.495	2.313	0.988	0.273
650				6.684	2.380	1.017	0.281
660				6.876	2.449	1.046	0.289
670				7.070	2.518	1.075	0.298
680				7.267	2.588	1.105	0.306
690				7.466	2.659	1.136	0.314
700				7.667	2.731	1.166	0.323
710				7.871	2.803	1.197	0.331
720				8.078	2.877	1.229	0.340
730				8.287	2.951	1.261	0.349
740				8.498	3.027	1.293	0.358
750				8.712	3.103	1.325	0.367
760				8.929	3.180	1.358	0.376
770				9.148	3.258	1.392	0.385

Table 2. Pressure loss in 2" to 8" Class 160 PVC pipe (continued).

	PRESSURE LOSS TABLES						
	HAZEN-WILLIAMS EQUATION (C = 150) CLASS 160 PVC						
	PRESSURE LOSS (PSI/100 FEET)						
			PIPE SIZE				
FLOW GPM	2"	2 1/2"	3"	4"	5"	6"	8"
775				9.258	3.297	1.408	0.390
800				9.819	3.497	1.494	0.413
825				10.394	3.702	1.581	0.438
850				10.985	3.912	1.671	0.462
875				11.591	4.128	1.763	0.488
900				12.212	4.349	1.858	0.514
925				12.848	4.575	1.954	0.541
950				13.498	4.807	2.053	0.568
975				14.163	5.044	2.154	0.596
1000				14.843	5.286	2.258	0.625
1025				15.538	5.533	2.364	0.654
1050				16.247	5.786	2.471	0.684
1075				16.970	6.044	2.582	0.714
1100				17.709	6.306	2.694	0.745
1125				18.461	6.574	2.808	0.777
1150				19.228	6.848	2.925	0.809
1175				20.009	7.126	3.044	0.842
1200				20.805	7.409	3.165	0.876
1225				21.615	7.698	3.288	0.910
1250				22.439	7.991	3.413	0.945
1275				23.277	8.290	3.541	0.980
1300				24.129	8.593	3.671	1.016
1325				24.996	8.902	3.802	1.052
1350				25.876	9.215	3.936	1.089
1375				26.771	9.534	4.072	1.127
1400				27.679	9.857	4.211	1.165
1425				28.602	10.186	4.351	1.204
1450				29.538	10.519	4.493	1.243
1475				30.488	10.857	4.638	1.283
1500				31.452	11.201	4.784	1.324
1550				33.421	11.902	5.084	1.407
1600				35.445	12.623	5.392	1.492
1650				37.524	13.363	5.708	1.580
1700				39.657	14.123	6.032	1.669
1750				41.844	14.902	6.365	1.761
1800				44.085	15.700	6.706	1.856

Table 3. Pressure loss in 1/2" to 2" Schedule 40 PVC pipe.

		PRESSURE LOSS TABLES				
		HAZEN-WILLIAMS EQUATION (C = 140) SCHEDULE 40 PVC				
		PRESSURE LOSS (PSI/100 FEET)				
			PIPE SIZE			
FLOW	1/2"	3/4"	1"	1 1/4"	1 1/2"	2"
GPM						
0.1	0.005					
0.2	0.018					
0.3	0.039					
0.4	0.066	0.023				
0.5	0.100	0.034				
0.6	0.139	0.048				
0.7	0.186	0.064				
0.8	0.238	0.082	0.025			
0.9	0.296	0.102	0.031			
1	0.359	0.124	0.038			
1.1	0.429	0.148	0.046			
1.2	0.504	0.173	0.054			
1.3	0.584	0.201	0.062			
1.4	0.670	0.231	0.071			
1.5	0.761	0.262	0.081	0.021		
1.6	0.858	0.295	0.091	0.024		
1.7	0.960	0.331	0.102	0.027		
1.8	1.067	0.367	0.113	0.030		
1.9	1.179	0.406	0.125	0.033		
2	1.297	0.447	0.138	0.036		
2.1	1.419	0.489	0.151	0.040		
2.2	1.547	0.533	0.164	0.043		
2.3	1.680	0.578	0.179	0.047	0.022	
2.4	1.818	0.626	0.193	0.051	0.024	
2.5	1.960	0.675	0.208	0.055	0.026	
2.6	2.108	0.726	0.224	0.059	0.028	
2.7	2.261	0.779	0.240	0.063	0.030	
2.8	2.418	0.833	0.257	0.068	0.032	
2.9	2.581	0.889	0.274	0.072	0.034	
3	2.748	0.946	0.292	0.077	0.036	
3.1	2.920	1.006	0.310	0.082	0.039	
3.2	3.097	1.066	0.329	0.087	0.041	
3.3	3.278	1.129	0.348	0.092	0.043	
3.4	3.465	1.193	0.368	0.097	0.046	
3.5	3.656	1.259	0.388	0.102	0.048	
3.6	3.852	1.326	0.409	0.108	0.051	
3.7	4.052	1.395	0.431	0.113	0.053	

Table 3. Pressure loss in 1/2" to 2" Schedule 40 PVC pipe (continued).

	PRESSURE LOSS TABLES				
	HAZEN-WILLIAMS EQUATION (C = 140) SCHEDULE 40 PVC				
	PRESSURE LOSS (PSI/100 FEET)				
			PIPE SIZE		
FLOW	1/2"	3/4"	1"	1 1/4"	1 1/2"	2"
GPM						
3.8	4.257	1.466	0.452	0.119	0.056	
3.9	4.467	1.538	0.475	0.125	0.059	
4	4.681	1.612	0.497	0.131	0.062	
4.1	4.901	1.688	0.521	0.137	0.065	
4.2	5.124	1.765	0.545	0.143	0.068	0.020
4.3	5.352	1.843	0.569	0.150	0.071	0.021
4.4	5.585	1.923	0.594	0.156	0.074	0.022
4.5	5.823	2.005	0.619	0.163	0.077	0.023
4.6	6.065	2.088	0.644	0.169	0.080	0.024
4.7	6.311	2.173	0.671	0.176	0.083	0.025
4.8	6.562	2.260	0.697	0.183	0.087	0.026
5.25	7.746	2.668	0.823	0.217	0.102	0.030
5.5	8.443	2.908	0.897	0.236	0.111	0.033
5.75		3.157	0.974	0.256	0.121	0.036
6		3.416	1.054	0.277	0.131	0.039
6.25		3.684	1.137	0.299	0.141	0.042
6.5		3.962	1.223	0.322	0.152	0.045
6.75		4.249	1.311	0.345	0.163	0.048
7		4.545	1.402	0.369	0.174	0.052
7.25		4.850	1.497	0.394	0.186	0.055
7.5		5.164	1.594	0.419	0.198	0.059
7.75		5.487	1.693	0.445	0.210	0.062
8		5.820	1.796	0.472	0.223	0.066
8.25		6.161	1.901	0.500	0.236	0.070
8.5		6.511	2.009	0.528	0.249	0.074
8.75		6.870	2.120	0.558	0.263	0.078
9		7.238	2.234	0.587	0.277	0.082
9.25		7.615	2.350	0.618	0.292	0.086
9.5		8.001	2.469	0.649	0.307	0.091
9.75		8.395	2.591	0.681	0.322	0.095
10		8.798	2.715	0.714	0.337	0.100
10.5			2.972	0.782	0.369	0.109
11			3.239	0.852	0.402	0.119
11.5			3.517	0.925	0.437	0.129
12			3.806	1.001	0.472	0.140
12.5			4.104	1.079	0.510	0.151

Table 3. Pressure loss in 1/2" to 2" Schedule 40 PVC pipe (continued).

			PRESSURE LOSS TABLES			
	HAZEN-WILLIAMS EQUATION (C = 140) SCHEDULE 40 PVC					
	PRESSURE LOSS (PSI/100 FEET)					
			PIPE SIZE			
FLOW	1/2"	3/4"	1"	1 1/4"	1 1/2"	2"
GPM						
13			4.414	1.161	0.548	0.162
12.5			4.104	1.079	0.510	0.151
13			4.414	1.161	0.548	0.162
13.5			4.733	1.245	0.588	0.174
14			5.063	1.332	0.629	0.186
14.5			5.403	1.421	0.671	0.199
15			5.753	1.513	0.714	0.212
15.5			6.113	1.608	0.759	0.225
16			6.483	1.705	0.805	0.238
16.5			6.864	1.805	0.852	0.252
17			7.254	1.908	0.901	0.267
17.5			7.654	2.013	0.950	0.281
18			8.064	2.121	1.001	0.296
18.5			8.483	2.231	1.053	0.312
19			8.913	2.344	1.107	0.328
19.5				2.460	1.161	0.344
20				2.578	1.217	0.360
20.5				2.698	1.274	0.377
21				2.822	1.332	0.394
21.5				2.947	1.391	0.412
22				3.075	1.452	0.430
22.5				3.206	1.513	0.448
23				3.339	1.576	0.467
23.5				3.475	1.640	0.486
24				3.613	1.706	0.505
24.5				3.754	1.772	0.525
25				3.897	1.839	0.545
26				4.191	1.978	0.586
27				4.494	2.121	0.628
28				4.807	2.269	0.672
29				5.130	2.421	0.717
30				5.462	2.578	0.764
32				6.156	2.906	0.861
34				6.887	3.251	0.963
36				7.656	3.614	1.070
38				8.462	3.995	1.183
40				9.306	4.393	1.301

Table 4. Pressure loss in 2" to 8" Schedule 40 PVC pipe.

			PRESSURE LOSS TABLES				
		HAZEN-WILLIAMS EQUATION (C = 150) SCHEDULE 40 PVC					
		PRESSURE LOSS (PSI/100 FEET)					
			PIPE SIZE				
FLOW	2"	2 1/2"	3"	4"	5"	6"	8"
GPM							
10	0.088						
15	0.186						
20	0.317	0.133					
25	0.479	0.202					
30	0.672	0.283	0.098				
35	0.894	0.376	0.131				
40	1.145	0.482	0.167				
45	1.424	0.599	0.208				
50	1.731	0.728	0.253				
55	2.065	0.869	0.302				
60	2.426	1.021	0.354	0.094			
65	2.814	1.184	0.411	0.109			
70	3.228	1.358	0.472	0.126			
75	3.667	1.543	0.536	0.143			
80	4.133	1.739	0.604	0.161			
85	4.624	1.946	0.676	0.180			
90	5.140	2.163	0.751	0.200			
95	5.682	2.391	0.830	0.221			
100	6.248	2.630	0.913	0.243			
105	6.839	2.878	0.999	0.266	0.088		
110	7.454	3.137	1.089	0.290	0.096		
115	8.094	3.406	1.183	0.315	0.105		
120	8.758	3.686	1.280	0.341	0.113		
125	9.446	3.975	1.380	0.367	0.122		
130	10.157	4.275	1.484	0.395	0.131		
135	10.892	4.584	1.592	0.424	0.141		
140	11.651	4.904	1.703	0.453	0.151	0.062	
145	12.434	5.233	1.817	0.484	0.161	0.066	
150	13.239	5.572	1.935	0.515	0.171	0.070	
155	14.068	5.921	2.056	0.547	0.182	0.074	
160	14.920	6.279	2.180	0.580	0.193	0.079	
165	15.795	6.648	2.308	0.614	0.204	0.084	
170	16.693	7.025	2.439	0.649	0.216	0.088	0.023
175	17.614	7.413	2.574	0.685	0.228	0.093	0.024
180	18.557	7.810	2.712	0.722	0.240	0.098	0.026
185	19.523	8.216	2.853	0.759	0.253	0.103	0.027
190	20.512	8.632	2.997	0.798	0.265	0.108	0.028

Table 4. Pressure loss in 2" to 8" Schedule 40 PVC pipe (continued).

			PRESSURE LOSS TABLES				
	HAZEN-WILLIAMS EQUATION (C = 150) SCHEDULE 40 PVC						
			PRESSURE LOSS (PSI/100 FEET)				
			PIPE SIZE				
FLOW	2"	2 1/2"	3"	4"	5"	6"	8"
GPM							
195	21.522	9.058	3.145	0.837	0.279	0.114	0.030
200	22.556	9.493	3.296	0.877	0.292	0.119	0.031
205	23.611	9.937	3.450	0.919	0.306	0.125	0.033
210	24.689	10.390	3.608	0.960	0.319	0.131	0.034
215	25.788	10.853	3.768	1.003	0.334	0.136	0.036
220		11.325	3.932	1.047	0.348	0.142	0.037
225		11.806	4.099	1.091	0.363	0.148	0.039
230		12.297	4.270	1.137	0.378	0.155	0.041
235		12.797	4.443	1.183	0.393	0.161	0.042
240		13.305	4.620	1.230	0.409	0.167	0.044
245		13.823	4.800	1.278	0.425	0.174	0.046
250		14.350	4.983	1.326	0.441	0.180	0.047
255		14.886	5.169	1.376	0.458	0.187	0.049
260		15.432	5.358	1.426	0.474	0.194	0.051
265		15.986	5.550	1.478	0.492	0.201	0.053
270		16.549	5.746	1.530	0.509	0.208	0.055
275		17.121	5.944	1.583	0.526	0.215	0.057
280		17.702	6.146	1.636	0.544	0.222	0.058
285		18.292	6.351	1.691	0.562	0.230	0.060
290		18.890	6.559	1.746	0.581	0.237	0.062
295		19.498	6.770	1.802	0.600	0.245	0.064
300		20.114	6.984	1.859	0.618	0.253	0.066
305		20.740	7.201	1.917	0.638	0.261	0.068
310		21.374	7.421	1.976	0.657	0.269	0.071
315		22.017	7.644	2.035	0.677	0.277	0.073
320		22.668	7.871	2.095	0.697	0.285	0.075
325		23.328	8.100	2.156	0.717	0.293	0.077
330		23.998	8.332	2.218	0.738	0.302	0.079
335		24.675	8.567	2.281	0.759	0.310	0.081
340		25.362	8.806	2.344	0.780	0.319	0.084
345			9.047	2.409	0.801	0.327	0.086
350			9.291	2.474	0.823	0.336	0.088
360			9.789	2.606	0.867	0.354	0.093
370			10.299	2.742	0.912	0.373	0.098
380			10.820	2.881	0.958	0.392	0.103
390			11.353	3.023	1.005	0.411	0.108
400			11.898	3.168	1.054	0.431	0.113

Table 4. Pressure loss in 2" to 8" Schedule 40 PVC pipe (continued).

PRESSURE LOSS TABLES							
HAZEN-WILLIAMS EQUATION (C = 150) SCHEDULE 40 PVC							
PRESSURE LOSS (PSI/100 FEET)							
PIPE SIZE							
FLOW GPM	2"	2 1/2"	3"	4"	5"	6"	8"
410			12.455	3.316	1.103	0.451	0.118
420			13.024	3.467	1.153	0.471	0.124
430			13.604	3.622	1.205	0.492	0.129
440			14.195	3.779	1.257	0.514	0.135
450			14.799	3.940	1.310	0.536	0.141
460			15.413	4.103	1.365	0.558	0.147
470			16.040	4.270	1.420	0.580	0.152
480			16.677	4.440	1.477	0.604	0.159
490			17.327	4.613	1.534	0.627	0.165
500			17.987	4.789	1.593	0.651	0.171
510			18.659	4.968	1.652	0.675	0.177
520			19.342	5.149	1.713	0.700	0.184
530			20.037	5.334	1.774	0.725	0.190
540			20.743	5.522	1.837	0.751	0.197
550			21.460	5.713	1.900	0.777	0.204
560			22.188	5.907	1.965	0.803	0.211
570			22.927	6.104	2.030	0.830	0.218
580			23.678	6.304	2.097	0.857	0.225
590			24.439	6.506	2.164	0.884	0.232
600			25.212	6.712	2.233	0.912	0.240
610			25.996	6.921	2.302	0.941	0.247
620				7.132	2.372	0.970	0.255
630				7.347	2.444	0.999	0.262
640				7.564	2.516	1.028	0.270
650				7.785	2.589	1.058	0.278
660				8.008	2.664	1.089	0.286
670				8.234	2.739	1.119	0.294
680				8.463	2.815	1.150	0.302
690				8.695	2.892	1.182	0.310
700				8.930	2.970	1.214	0.319
710				9.168	3.049	1.246	0.327
720				9.408	3.129	1.279	0.336
730				9.652	3.210	1.312	0.345
740				9.898	3.292	1.346	0.353
750				10.147	3.375	1.379	0.362
760				10.399	3.459	1.414	0.371
770				10.654	3.544	1.448	0.380

Table 4. Pressure loss in 2" to 8" Schedule 40 PVC pipe (continued).

			PRESSURE LOSS TABLES			
	HAZEN-WILLIAMS EQUATION (C = 150) SCHEDULE 40 PVC					
			PRESSURE LOSS (PSI/100 FEET)			
			PIPE SIZE			
FLOW	2"	2 1/2"	3"	4"	5"	6"	8"
GPM						
775				10.782	3.587	1.466	0.385
800				11.435	3.804	1.555	0.408
825				12.106	4.027	1.646	0.432
850				12.794	4.256	1.739	0.457
875				13.500	4.490	1.835	0.482
900				14.223	4.731	1.933	0.508
925				14.963	4.977	2.034	0.534
950				15.720	5.229	2.137	0.561
975				16.495	5.487	2.242	0.589
1000				17.287	5.750	2.350	0.617
1025				18.096	6.019	2.460	0.646
1050				18.922	6.294	2.572	0.676
1075				19.765	6.574	2.687	0.706
1100				20.624	6.860	2.804	0.736
1125				21.501	7.152	2.923	0.768
1150				22.394	7.449	3.044	0.800
1175				23.304	7.752	3.168	0.832
1200				24.231	8.060	3.294	0.865
1225				25.174	8.374	3.422	0.899
1250				26.134	8.693	3.553	0.933
1275				27.110	9.018	3.685	0.968
1300				28.102	9.348	3.820	1.003
1325				29.112	9.683	3.957	1.039
1350				30.137	10.024	4.097	1.076
1375				31.179	10.371	4.238	1.113
1400				32.237	10.723	4.382	1.151
1425				33.311	11.080	4.528	1.189
1450				34.401	11.443	4.677	1.228
1475				35.508	11.811	4.827	1.268
1500				36.630	12.184	4.980	1.308
1550				38.924	12.947	5.291	1.390
1600				41.281	13.731	5.612	1.474
1650				43.702	14.537	5.941	1.560
1700				46.186	15.363	6.279	1.649
1750				48.733	16.210	6.625	1.740
1800				51.344	17.078	6.980	1.833

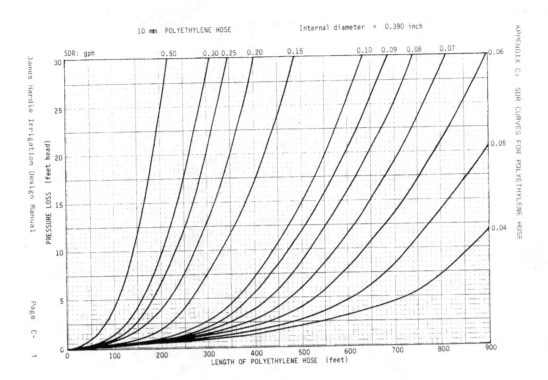

Figure 1. Pressure loss curves for 10 mm polyethylene tubing.

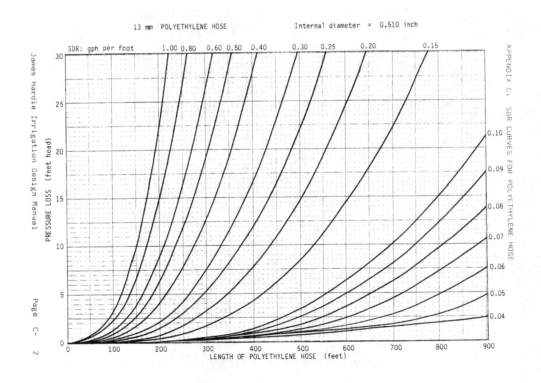

Figure 2. Pressure loss curves for 13 mm polyethylene tubing.

Charts courtesy of Hardie Irrigation, El Cajon, CA 92022-2246

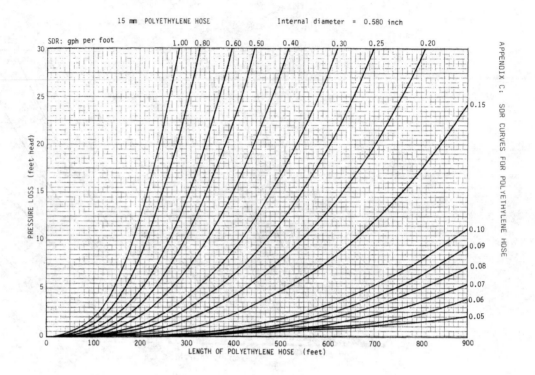

Figure 3. Pressure loss curves for 15 mm polyethylene tubing.

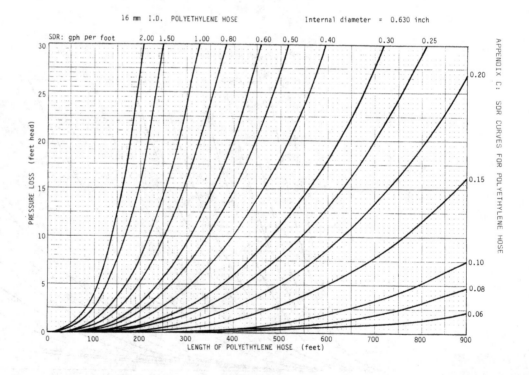

Figure 4. Pressure loss curves for 16 mm inside-diameter polyethylene tubing.

Charts courtesy of Hardie Irrigation, El Cajon, CA 92022-2246

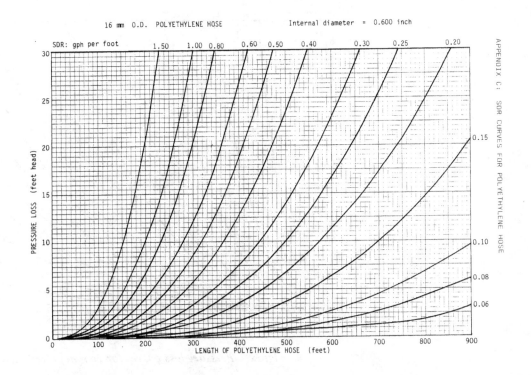

Figure 5. Pressure loss curves for 16 mm outside-diameter polyethylene tubing.

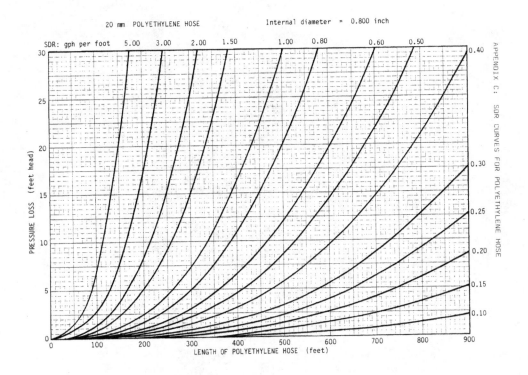

Figure 6. Pressure loss curves for 20 mm polyethylene tubing.

Charts courtesy of Hardie Irrigation, El Cajon, CA 92022-2246

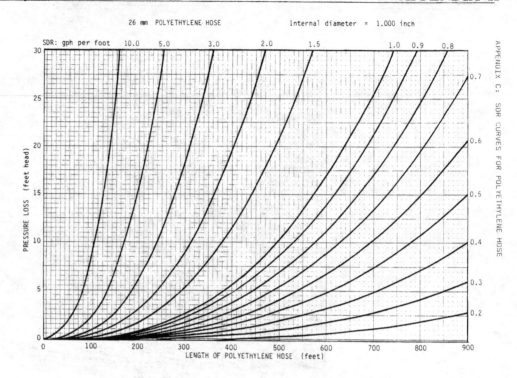

Figure 7. Pressure loss curves for 26 mm polyethylene tubing.

Glossary

Acidification. Adding sulfuric acid, phosphoric acid, or hydrochloric acid to a micro-irrigation system in order to lower the water pH to counteract chemical precipitation, which can cause clogging.

Amendment. See *Soil amendment;* see *Water amendment.*

Application efficiency (AE). The efficiency of a single irrigation event, calculated as the water stored in the crop root zone, divided by the total water applied, multiplied by 100.

Backwashing. Cleaning filters by reversing the flow of water or discharging water over the screen to wash away contaminants.

Batch Tank. See *Differential pressure tank.*

Chemical precipitation. Process whereby chemicals present in water separate from solution and become deposited in the irrigation system, which can clog drip emitters.

Chemigation. Injecting chemicals through an irrigation system

Chlorination. Adding chlorine to an irrigation system to counteract clogging caused by organic materials such as algae, moss, and bacterial slimes.

Coefficient of variation (CV). A measurement of the difference in output among emitters resulting from the manufacturing process. This measurement is provided by the manufacturer. The standard deviation divided by the mean value of a sampling of emitters.

Crop water use. The amount of water used by a crop in a given time period. See also *Evapotranspiration.*

Deep percolation. Process in which irrigation water flows through the soil down past the root zone and is therefore no longer available to the crop.

Dielectric constant meter. Device for measuring soil moisture.

Differential pressure tank. A device for injecting chemicals through an irrigation system. The tank inlet is connected to the irrigation system at a point of pressure higher than that of the outlet connection, causing irrigation water to flow through the tank containing the chemical to be injected. Also called a "batch tank."

Discharge rate. The rate at which water is discharged from emitters, expressed as a volume per unit of time — either gallons per hour or liters per hour.

Distribution uniformity (DU). A measure of how uniformly water is applied over a field, calculated as the minimum depth of applied water (usually the average discharge of the lowest 25% of all emitters measured), divided by the average discharge of all emitters measured, multiplied by 100. Sometimes referred to as **Emission uniformity.**

Double-chambered tape/tubing. Drip tape or tubing consisting of a large main chamber, which conveys the water along the lateral, and a small secondary chamber with orifices in the outer wall through which water is discharged into the soil.

Emission uniformity. See **Distribution uniformity.**

Emitter. Device through which water is discharged to the soil surface from the lateral lines in a micro-irrigation system. Emitters may be drip emitters, line sources, or micro-sprinklers.

Emitter discharge coefficient. The relationship between the dimensions and configurations of flow passages and consequent pressure loss through the emitter.

Emitter discharge exponent. The relative ability of an emitter to compensate for variations in pressure. The smaller the exponent, the less sensitive the emitter discharge rate to pressure variation.

Evapotranspiration. The amount of water used by a particular type of crop in a given period of time, comprised of water evaporating from the soil and water transpiring from the plants. Crop evapotranspiration estimates are available as either historical averages or real-time estimates from the California Department of Water Resources CIMIS program and University of California Cooperative Extension offices.

Fertigation. Applying liquid fertilizer through an irrigation system.

Filtration. The removal of suspended solids from irrigation water.

Friction loss. A loss of pressure caused by friction as water moves through the irrigation system.

Gross irrigation. The amount of irrigation water applied to satisfy the needs of the crop and to compensate for inefficiency and non-uniformity in the irrigation system. Calculated as the net irrigation amount, divided by the application efficiency, multiplied by 100.

Gypsum block. *See* *Resistance block.*

In-line emitters. Drip emitters installed inside the drip tubing as part of the tubing flow path.

Irrigation efficiency (IE). A measure of the portion of total applied irrigation water beneficially used — as for crop water needs, frost protection, salt leaching, and chemical application — over the course of a season. Calculated as beneficially used water, divided by total water applied, multiplied by 100.

Lateral line. The water delivery pipeline or polyethylene hose that supplies water to the emitters from the mainlines or submains.

Leaching. Applying irrigation water in excess of the soil moisture depletion level to remove salts from the root zone.

Leaching fraction. The percent of infiltrated water that percolates below the root zone.

Line-source. Type of low-volume emitter consisting of drip tape or tubing with emission points at regular intervals.

Micro-irrigation. Applying irrigation water through a system of tubing or lateral lines and low-volume emitters such as drippers, bubblers, drip tapes, or micro-sprinklers. Also called *drip irrigation* or *trickle irrigation.*

Mainlines and submains. The water delivery pipelines that supply water from the control head to the laterals.

Mazzei injector. A brand of venturi device for injecting chemicals through an irrigation system. The device consists of a constriction in the pipe flow area, resulting in a negative pressure or suction at the throat of the constriction.

Media filter. An irrigation water filter that removes suspended solids by passing water through sand.

Net irrigation amount. The amount of water used by the crop since the last irrigation.

Neutron probe. Device for measuring soil moisture using a radioactive source.

pH. A measure of the acidity or alkalinity of a liquid. A pH of 7.0 is neutral; a pH less than 7.0 is acidic; and a pH greater than 7.0 is alkaline.

Positive displacement pump. Piston or diaphragm pumps, powered by electricity, water, or gasoline, for injecting chemicals through an irrigation system at precise rates.

Propeller flowmeter. Device for measuring flowrates in pipelines, consisting of a propeller linked by a cable or shafts and gears to a flow indicator.

Resistance block. A device, usually made of gypsum, that measures the conductivity of current passing between two electrodes to indirectly measure soil moisture.

Salinity. A measure of the soluble salts in a soil or water. Usually measured as electrical conductivity (EC) in millimhos per centimeter (mmhos/cm) or deciSiemens per meter (dS/m).

Sand separator. Filter used to remove large heavy particles from water.

Single-chambered tape/tubing. Drip tape or tubing consisting of a single main chamber. Water flows from this chamber into drip emitters spaced along the tape or tubing and is discharged from the emitters into the soil.

Soil amendment. A substance added to the soil primarily to improve the soil's physical condition.

Specific discharge rate (SDR). The emitter discharge rate divided by the emitter spacing.

Strip emitter. Type of emitter suitable for use with single-chambered tape or tubing. Strip emitters are long and narrow in shape and are usually classified as turbulent-flow emitters because the configuration of the flow path creates turbulence in the water, allowing for larger flow passages, which can reduce the potential for clogging.

Surface runoff. Water flowing off the surface of the field.

Tensiometer. A device for measuring soil moisture

Uniformity. *See* **_Distribution uniformity_**

Venturi Device. Device for injecting chemicals through an irrigation system. The device consists of a constriction in the pipe flow area, resulting in a negative pressure or suction at the throat of the constriction. *See also* **_Mazzei injector._**

Water amendment. A chemical added to water to improve soil-water properties, such as water infiltration.

Index